SOUTHERN STELLAR OBJECTS FOR SMALL TELESCOPES

SOUTHERN STELLAR OBJECTS FOR SMALL TELESCOPES

BETWEEN THE EQUATOR AND 55 SOUTH DECLINATION, WITH OBSERVATIONS MADE IN THE PUNJAB

JOHN ELLARD GORE

GYAN

Published by

GYAN

Gyan Publishing House
5, Ansari Road
Daryaganj, New Delhi-110002
Phone: 011-47034999, 9811692060
E-mail: books@gyanbooks.com

Distribution Network
gyanbooks.com
India, USA, Canada, UK, Australia

ISBN: 978-93-7447-469-3 (PB)

First Published, 1877

2nd Impression 2026

Printed at: Gyan Press, Delhi.

SOUTHERN STELLAR OBJECTS FOR SMALL TELESCOPES
Author: JOHN ELLARD GORE

SOUTHERN STELLAR OBJECTS.

SOUTHERN STELLAR OBJECTS

FOR SMALL TELESCOPES,

BETWEEN THE EQUATOR AND 55° SOUTH DECLINATION,

WITH OBSERVATIONS MADE IN THE PUNJAB,

BY

J. E. GORE, M.R.I.A. A.I.C.E.,

ASST. ENGINEER, PUBLIC WORKS, INDIA.

LODIANA:

PRINTED AT THE LODIANA MISSION PRESS;

1877.

PREFACE.

The objects in the following pages have been chiefly selected from Sir John Herschel's great work "*Results of Astronomical Observations at the Cape of Good Hope*", and are intended for telescopes from 2 to 4 inches in aperture. The original observations given were made in the Punjab with an achromatic of 3 inches aperture and 4 feet focal length made by Mr. Browning, London; and an object glass of 3.9 inches clear aperture and 5 feet focus by the eminent maker Wray, mounted on a simple equatorial stand* by Clarkson, London. These observations are—as in Mr. Webb's *Celestial Objects*—included in brackets thus [ ]. Observations on some of the objects observed by Mr. Webb are also given. Other remarkable objects noticed in searching for the doubles and nebulæ in Sir J. Herschel's Catalogues are alluded to in passing, and it is hoped that the list will be found useful to amateur observers in latitudes South of England. Many of the objects however can be observed in England, though most of them of course at a small altitude above the Southern horizon.

The leading number is that of the star or nebulæ in Sir J. Herschel's Catalogues of Southern doubles and nebulæ. In many cases the number of the object in other observer's Catalogues is also given.

---

* Without circles.

The R.A. and Declination of the objects have been brought up to 1880.

The quotations are from Sir J. Herschel's very graphic description of the objects he observed.

As in '*Celestial Objects*', the magnitudes of the stars composing a double or triple are given after the R.A and Declination; next the position angles, and lastly the distance in seconds.

The 3 inch telescope was mounted, as an alta-zimuth on a round brick pillar about 5½ feet high, and as a cheap and very steady mounting I can recommend the plan to amateurs, who can manage to have a pillar of the sort built in an open situation.

As in the Cape Observations,

A. stands for the Royal Astronomical Society's Catalogue of stars for 1830.

B. The Brisbane Catalogue.

L. Lacaille's Catalogue of 9766 stars in the Southern Hemisphere, published by the British Association. The star places for 1750.

LL. Lalande's Catalogue of 47,390 stars from the *Histoire Céleste Française*, published by the British Association. The star places for 1800.

B.A.C. The British Association's Catalogue of stars.

J. E. G.

NABHA, PUNJAB, *January* 1877.

SOUTHERN STELLAR OBJECTS.

ANTLIA.

A small constellation, situated between Argo and Centaurus, and South of Hydra. It is particularly devoid of conspicuous stars, the brightest being χ, a small star of the 4th magnitude. The following objects, however, are deserving of notice.

DOUBLE STARS, ETC.

4271. η. 9^h. 54^m, *S* 35° 19′; 4, 13 : 312°.1 : 30″. Herschel also gives 6, 13 : 314°.7 : 25″, and says 'Large star single with 180'. It was rated 6m. by Lacaille (4095).

4321. δ. 10^h 24^m, *S* 29° 59′; 6, 10 : 225° ± : 10″. 'Very neat star.'

4216. ζ. $9^h\ 25^m$, S 31° 12′ (1870); 6, 7; 210°. 6 : 8″.38 (1836). [Smaller star bluish, 3 inches, 1873. It is the *preceding* of two 6m. stars close together, *South following* • Mali].

Red star. 3630. B. A. C. $10^h\ 30^m$, S 38° 54′; $6\frac{1}{2}$m, ; *s. f η*; described by Sir J. Herschel as 'Extreme orange almost scarlet.' It seems to be identical with Lacaille 4358 (7m).

AQUARIUS.

A large constellation forming one of the signs of the Zodiac. It lies south of Equulens, Pegasus and Pisces, and North of Capricornus and Piscis Australis. It contains no star brighter than the 3rd mag. but includes within its boundaries many interesting telescopic objects.

Double stars, etc.

5363. Struve 2944; $22^h.\ 42^m$ S 4°. 49′; 7, 7 : 345°.3 : 4″ 'A third 10m star distant about 60″. Marked in Proctor's Atlas as a suspected Trinary.

About 2°½ following χ Aquarii . Smyth gives 7½, 8, 9 : 247°.4 , 158° ; 4″.2 , 55″.1 . Dembowski found in 1862 , 146°.7 and 50″.7, which would seem to show a considerable change of position in the distant companion.

5389 . ψ.[1] 23ʰ. 10ᵐ, *S* 9°. 44′, 5½, 10 : 311° : 49″. 3. Webb gives 5.5, 9 , and the colours, orange or topaz yellow, sky blue [*Comes* seen with 40 on 3 in., in strong moonlight , 1875 , 1876 . A faint and distant *comes* about 11m. *n. p.* A little *n. p.* is a wide pair, D about 30″ . P. 170°±, with a 3rd star *s. f.*]

Struve 3008. 69. P. XXIII. 23ʰ. 18ᵐ, *S* 9°. 5′ ; 7½, 8 : 90° : 6″.88 (1837) . Smyth gives 8, 8½ : 272°.1 : 7″ .5 (1834). [A little *n. f.* ψ[2] and ψ[3] Aquarii. Well seen with 83 on 3 in. Nov. 1876. The distance seems rather less than 7″. The *f* star is certainly the brighter of the two—nearly 1 mag. difference. Herschel would seem to have seen the *preceding* star the brighter as he has made it his primary. Can it be variable ?] Herschel says " Exactly in the parallel so as to allow of no motion of the wire without destroying the measure. Most carefully examined in every way". Smyth remarks that " A has a proper motion, which supposing B to remain stationary, would in half a century, increase the angle to 340°, and diminish the distance to about 5″. [So large a change in angle has certainly not taken place, as the companion is still not far from the parallel—as seen by Herschel and Smyth. The angle seems to have diminished, not increased. The object is preceded by 2 small stars about 9m, forming with it an isosceles triangle. A line joining the pair would pass between these two stars Nov. 1876]. The pair is now believed to be a binary one.

5397. 23^h 18^m, *S* 15° 9′ . 7, 9 : 330° : 62″ (1835). [There is a fainter star to the South nearly in a line with the pair, 3 inches, 1876].

τ′. 22^h. 41^m, *S* 14° 41′ . 6, 9½ : 113°.5 : 29″.7 (Smyth 1838). [*Comes* very small with 3 inches, Nov. 1876. More like 11m. Certainly fainter than the companion to Polaris].

In 1874, the colours were noted by H Sadler

as "greenish white and amethyst", with a 6½ inch silver-on-glass reflector.

ARGO

This large constellation has been divided into four divisions called Carina, Malus, Puppis, and Vela. In these pages however, the whole will be considered as one constellation, the division to which any star belongs being stated where necessary.

Double stars, etc.

α. *(Canopus)*. $6^h.21^m$, *S* 52°.38′. This fine star, second only to Sirius in brilliancy, does not rise above the English horizon, but is a conspicuous object in Northern India in the winter months. Webb says "It was thought, 1861 in Chili, brighter than Sirius". [Though attaining a meridian altitude of only 7° at my station in the Punjab, I have several times (1874) noticed it to be very little inferior to Sirius. It may be variable, and should be watched by Southern observers].

3883. 2193. B. A. C. V Puppis. $6^h. 35^m$, *S* 48°. 6′; 5,7 : 317°: 13″.4, pale yellow, pale blue, [7 more like 8, 3 in. 1875. Two *comites* about 10m. *n p* and *s.f* on opposite sides of the pair. A little *n.f* is a fine open pair 6m].

3884. Dunlop 32. B 1335. $6^h. 38^m$. *S* 38°. 17′; 6.5,8 : 276°.1 : 8″. 43 (1835). 'Fine star'. One of a stream of 6m stars from Canis Major to ν Puppis (3m).

3889. B 1340. $6^h. 40^m$, *S* 50°. 20′; 6½, 8½ : 266°.1 : 42″.46. (1836) [Near τ. Seen, but not well, as it is very low in the Panjab sky].

3895. $6^h. 42^m$, *S* 47°. 37′; 7, 11 : 61° : 25″, orange, blue [Not examined].

3928. 7^h, 1^m, *S* 35°. 35′. Triple; 6½, 8½, 10 : 157°.7, 286°.5 : 5″.39, 37″.04 (1836). Herschel says "The nearer star B bears great illumination, the farther C does not; 2 others near on opposite sides." [White, bluish, 10 more like 11 , 3 inches, 1875. A fainter and more distant *comes s. p.* makes it quadruple. About midway between π Argus and κ^2 Canis Majoris].

3931. $7^h. 3^m$, *S* 42°. 8′ ; 6½, 7½ : 40°.8 : 72″.92 (1836). Near *c* Puppis (5m) . One of a stream of 5m and 6m stars running from π, South between ν and σ . It is 2631 in Lacaille's Catalogue.

3946. π. 7^h 13^m. *S* 36°, 52′ ; 3, 8 : 211°.8 : 70″ ; yellow, blue [large star a beautiful orange, 3 inches, 1875].

3966. L 2801. $7^h. 21^m$, *S* 37°. 3.7, 7 : 140° 9 : 7″.03 (1835). About 1°½*f* π. [Both white or light yellow, 1875 . 40 just divided it with 3 inches].

3972 . σ Argus . $7^h. 25^m$, *S* 43°.4′ ; 5, 10 : 75° : 22″ .52 , Sir J. Herschel says " The small star bears a reasonable illumination, steady but ill defined . σ is 4m. in Proctor's Atlas and was rated 3m. by Lacaille in 1752. [10 seen with power 40 on 3 inches 1875, but not with 133. A faint and distant *comes, South preceding* ; another more distant to the South].

[A little North of a line joining σ with ζ is a star W shown in Proctor's Atlas as 4m. I noticed in March 1875 that this star was only 6 or 6½m, being then much fainter than the 5m star a between it and ζ. W was rated 4½m. by Lacaille in 1751, and may therefore possibly be variable].

3982. L 2867, $7^h. 31^m$, S 28°. 6′. 'Coarsely triple', 6, 9, 9 [Near η Canis Majoris . η is finely grouped with a 7m star, and several other fainter stars surrounding it].

P. VII, 149, 147 . 52 Dunlop. h^2 Puppis . $7^h. 29^m$, S 23°. 13′ ; 6½, 6½ : 106°.8 : 9″.37 . Webb gives from Smyth 6, 6 : 284°.1 : 10″.2 , and says '*n.p* rather larger 1851, 1856' ; so Herschel seems to have seen it as he made the *n.p.* star his primary [A splendid pair, both white, or very light yellow. The *n.p* star *very* slightly the brighter 1875].

P VII 175, 177 ; Dunlop 53. κ' Puppis. 7^h. 34^m, *S* 26°. 30′ (1870) ; 5, 5 : 137°.8 : 10″.26. Smyth gives both 6½ : 326.8 : 9″.8 (1831.9) but the magnitudes are evidently underrated, the object having been, as he supposes, mistaken for h^2 (above). It was rated 5m by Lacaille [A splendid pair ; both white or very light yellow, 3 in. 1875].

4028. B 1848. 7^h. 56^m, *S* 59° .40′ ; 7, 7 ; 46°.9 : 17″.05 (Not examined).

4030. 7^h. 57^m, *S* 40°. 58′ ; 7, 9 : 345° : 40.″ " Large star white, small rich ruby coloured, approaching to sanguine. Very remarkable" [A little *s. p* ζ. The more Southern and the brighter of two stars about 10′ apart. 9 small, more like 10m, 3in. 1875].

4049. γ Argus. B 1914. Dunlop 63. 8^h. 6^m, *S* 46°. 59′ : 2, 6, 8 : 220°.2, 151°.6 ; 41″.16, 62″.4. Herschel says " besides the two principal stars are three others 11, 11, 13 arranged in an exact right line". [A splendid open pair. The two smaller stars, about 8m and 9m. There are two more stars on opposite sides forming with the principal pair a curved line. γ is marked 3m in Proctor's Atlas, but was rated 2m by Lacaille in 1752. Sir J. Herschel at the Cape also estimated γ as about 2m, and brighter than λ; and so I found it in March 1875].

4058. B 1938. $8^h. 10^m$, *S* 35°. 32′; 6, 7: 190°±: 4″. 'Very ill defined' [Near *r* Puppis. From the position given, this star should be in the same low power field with Dunlop 67 (below). There *is* a 6m star in the place indicated, but with 133 on 3in. (1875) I saw the star single so that if really a double star it must be much closer than 4″].

4062. h^2 Puppis. $8^h.10^m$. *S* 39° 58′; 5. 5, 10: 343°.4: 50″; yellow, blue [Nearly *fς*. Seen].

About 4° North of this, are some fine fields with several wide pairs. Proctor shows a 5m and a 6m star South preceding *p* and *r*. [These I found about 7m in Feb. 1875].

67 Dunlop. B 1941, 1942. $8^h. 11^m$, *S* 35° 57′ ; 6, 7 : 175°.2 : 67″.45, 1837. Near *r* Puppis [Yellowish, bluish white, 1875. In a fine field. Preceding is a wider pair about 8m. South of these pairs is *h* 4060, 9, 10 : 21″, and further South is another small pair about 8½, 11, of which the larger star is red, 3 inches, 1875].

4063. $8^h. 11^m$, *S* 36° 58′. 8, 9½ : 346°.3 : 20″. 'Large star white, small star so very red as to give the large star by contrast a bluish tint'. Near *r* Puppis.

4104. $8^h. 25^m$, *S* 47° 31′. 6, 9, 10 : 242°.3, 39°.9 ; 4″.4, 20″. 'Beautifully triple'. This is A Velorum in Proctor's Atlas *s. f.* γ.

4115. L 3434. $8^h. 33^m$, *S* 33° 19′ ; 6.5, 12, 13 : 155°.6, 21°.3 : 20″, 40″. *S. p. a* Mali (4m).

4120. B 2122. $8^h. 35^m$, *S* 29° 7′. 5½, 11 : 40°± : 50″. This is *f* Mali [Well seen with 40 on

3 inches. About 25′ North is a wide pair, 7½, 11 : 80″± of which the large star is orange, 3 in. 1875].

4127. *b* Velorum. $8^h. 37^m$, *S* 46° 12′. 5, 11 : 57°.7 : 30″ [A little South of a line joining γ and λ (2½m). *Comes* not seen with 3 in., 1875].

4133. *d* Velorum. $8^h. 40^m$, *S* 42° 13′. 5, 11 : 60°.8 : 47″.2, yellow, blue [Between ζ and λ, nearer to the latter. *Comes* not seen with 3 inches, 1875 ; must be fainter than 11m which in such a wide pair should be plain with 3 in. (See 4120 above) as the primary is not a brilliant star. The object is however rather low in the Punjab sky].

4183. A 1115. • Mali. $9^h. 6^m$, *S* 29° 49′ ; 6½, 9½ : 144°.9 : 18″. Herschel also gives 7, 12 : 140°.5 : 15″ [9½ very small, more like 11m. The merest point with 40, 3 in. 1875].

76 Dunlop. B 2505. $9^h. 24^m$, *S* 55° 58′ ; 6½,

7 : 97°.8 : 59″.57, 1838. About 5° *s. f.* . It was rated 7m by Lacaille.

77 Dunlop. B 2512. 9^h. 26^m, *S* 44°. 1′ ; 8, 8 : 257°.8 : 108″.47, 1837. A little *n* of 76 Dunlop.

4330. B 3069. *t* Velorum ; 10^h. 28^m, *S* 46° 22′ ; 6, 9 : 164.4 : 40″.66 (1836) ; orange, blue. [About 4° *n. p.* μ Velorum (3m). 9 small, more like 10m, 3 in. 1875. A 7m star follows. About 2° North of this is *s* Velorum, 4328 *h* 6½, 6½, both white and almost exactly equal].

λ, 2½m, is an orange star in a fine field. Mr. Pope of Dunedin, New Zealand, calls this star " blood red"*.

Red star 8^h. 40^m, *S* 27° 44′. 'A fiery red' star 8½m. About 1° *n. f* the 5m star *f* Mali.

---

* English Mechanic, June 25, 1875.

Red star, ϱ Mali. 3121 B. A. C. $9^h. 2^m$, S 25° 20′, 4½m. Rated 6m by Lacaille in 1752. It is 5m in Harding's Atlas.

Red star. $9^h. 50^m$, S 40° 58′. 7. 5m. "Scarlet" remarkably full rich colour' (Herschel).

CLUSTERS, AND NEBULA.

3103. Dunlop 535 $7^h. 47^m$, S 38° 1′ (1870) Herschel says "Superb Cluster, gradually brighter in the middle; 20′ diameter. Rich. Stars very remarkably equal. All 12 or 13m. Very few 14m; none 11m". North preceding ζ Puppis and close to *b*. The brighter stars well seen with 133 on 3 in. (1875). A stream of larger stars runs from it in a Northerly direction like the stem of a cornucopia].

3106. ℌ VII. 10. $7^h. 51^m$, S 22° 4′. 'A very rich milky way cluster or mass of stars, 10, 11 and 12m'. A little *n. f.* ξ Puppis and between it and P.

3132. $\mathrm{H}_{\circ}$ VII. 63. $8^h. 32^m$, *S* 29° 32′. 'A fine large pretty much compressed cluster; irregularly elongated, 10′ × 7′ ; stars 12 and 13m nearly equal.' Closely *s. p. f:* Mali (5m).

3179. $9^h. 30^m$, *S* 46° 21′. "An enormous cluster of a degree and a half in diameter, very rich in stars of all magnitudes from 8m downwards; a sort of telescopic Præsape". A few degrees *s.f* λ Argus (2½m) and North of M Velorum.

3228. 10^h. 1m, *S* 39° 48′. 'Very large, very bright; Planetary nebula, a well defined elliptic disc, with a 9m. star near the centre; diam. in R. A. 4^s (=1′)". 'The star is sharp; the nebula *velvety* or like infinitely fine dust. Six stars 9½m in the field; and one companion=13m' A little *n. p. q.* Velorum (4½m).

CANIS MAJOR.

A constellation remarkable for containing Sirius, the brightest star in the heavens. The star γ may possibly be variable, as it was rated 3m by Lalande.

It is now certainly fainter. Heis rates it 4.5m.

South of Sirius is an irregular shaped crucifix formed by δ, ϵ, η, and 30 (5m).

Double stars, etc.

α (*Sirius*).

$6^h. 40^m$, *S* 16°. 31′.

This magnificent white star is the brightest in either hemisphere, though it has been on one occasion, nearly equalled by the variable η Argus, and also, it has been said by Canopus (which see). It is attended by a 10m star nearly following (1861) discovered by Alvan Clarke in 1861. The combination is a binary one. According to Dr. Auwers the period is 49.399 years, or rather this is the period assigned by him for a disturbing body, which would account for the observed anomalies in the proper motion of Sirius. Mr. Wentworth Erck estimated the position angle in 1876 at 55°. The *comes* was at its maximum distance, 10″.8, in 1871 and the distance is now diminishing.

Several other fainter and more distant *comites* have been detected by other observers with large telescopes. Sirius is supposed to have changed in colour since the days of Ptolemy, who is said to have compared it in colour with Antares. It has a considerable proper motion in a southerly direction, and is—according to the Spectroscopic researches of Dr. Huggins—receding from the earth at the rate of about 20 miles per second. Vogel however makes this velocity 46 miles!

3857. B 1229. $6^h. 20^m$, *S* 36° 38′. Triple. 5.5, 12, 7. 257°.6, 66.2 : 12″, 69″.54 [About 2°. *s.f* κ Columbæ (4m). 12 not seen, 3 in. 1875. There is an 11m *comes* about 2′½ distant, nearly due North of the pair].

3858. $6^h. 21^m$, *S* 33° 57′. 7½, 8 : 313°.8 : 3″. 'Fine object'. A star 6m precedes to the South. *s.f* δ (4m).

3871. L 2337. $6^h. 30^m$, *S* 29° 31′. 7.5, 8 : 353°.1 : 10″. 'A fine double star'. Between ε and ζ, nearer the latter; in a rather vacant space, near a 7m star.

3873 . ν' 6^h. 31^m, *S* 18° 34′ . 7 , 8½ : 260°.9 : 20″. Webb gives "6.5,8 : 17″ : pale garnet, grey". 3° *s. p Sirius* [8½ bluish, 3 inches, 1875].

3945 . 7^h. 11^m, *S* 23°.0′ . 6.5, 7.5 : 67°.6 : 28″.21 (1837) "orange and green, fine contrast of colours". In another place Herschel says 'very high yellow, contrasted blue' [red, blue, 3 in., 1875. Visible to naked eye, North of 30 ; magnitudes seem underrated ; more like 6 and 7 . It is Lalande 14400 (5m). There is a *comes* about 11mag. nearly *following* and 90″± distant from 6.5 . A fainter and more distant *comes* to the South. Other small stars visible in a low power field.]

17. π^2 . 6^h. 49^m, *S* 20° 15′, is an open quadruple star, 6, 9½, 10 , 11 : 45″, 52″, 125″ [An interesting object for a small telescope, as a low power shows it well. The stars are arranged roughly in the shape of a long crucifix. A stream of small stars 11m and 12m runs from the 11m star towards the West.]

3948 . A 905 . 30 Canis Majoris . 7^h. 14^m,

S 24°. 43′. Triple 5½, 11, 12 : 85°.8, 73°.3 : 8″, 15″. 'Chief of a fine cluster'. Webb gives from Smyth 6.5, 9 : 73° : 85″, so that Smyth apparently did not see the closer companions. Burnham considers 11 and 12 to be fainter, as they are not very easy objects with a 6 inch refractor. The companions probably belong to the cluster, which may possibly be much farther from us than the bright star. [With 3 in I see 2 or 3 small stars rather nearer than Smyth's *comes.*]

About 1°½ *n. p* 30, Harding shows a 6m star. In Feb. 1875, I observed this star to be only 8 or 8½m. About 8½m, Jan. 1877.

η. 7^h. 20m, S 29° 5′. 3, 7½ : 285° : 169″ ; 'pale red dull grey ; two small stars following' (Smyth 1833) [Seen. Large star *not* red with 3 inches 1877].

β. R. A. 6^h. 17^m, S 17°. 54′. 2m Lalande and Harding 3. 2 Heis. It is now certainly fainter than δ and ε which Harding marks as 3m. Sir J. Herschel at the Cape estimated β at 2.58, δ being 2.32 and ε 1.86 on the same scale.

I found β very slightly superior to κ Orionis in

Jan. 1876. Smyth (Cel. Cycle) mentions a "dusky grey" 9m companion *n. p.* "and another of the same mag. in the *s. p.* quadrant."

δ. RA. $7^h. 3^m$, *S* 26°. 12′. 2m Lacaille, 3m Harding, 2m Proctor, 2m Heis, Sir John Herschel 2.32.

. It seemed to me rather less than 2m, but certainly brighter than 3m. Jan. 1876. Smyth (Cel. Cycle) says "It is considered variable; having been registered 2nd Mag. by Hevelius, Lacaille and Brisbane; 2½ by Halley; 3 by Ptolemy, Tycho and Flamsteed, and 3½ by Piazzi and Johnson". Smyth accepts 3½m as correct, but it is—now at least—most certainly brighter than this.

ε. RA. $6^h. 54^m$, *S* 28°. 48′, 3m Lacaille and Harding. 2.1 Heis, Sir J. Herschel 1.86. Its low meridian altitude in the sky of central Europe may account for its being underrated by Harding, but it is not so easy to understand why it should have been rated 3m by Lalande who observed it at the Cape of Good Hope, if it was of the same brilliancy in 1752 as at present.

[Certainly 2m., but not very much brighter than δ, Jan. 1876].

Harding shows an 8m star closely following. Smyth gives this 7m.

There is a close *comes* about 7″ distant which Burnham finds easy with 6 in, but which Smyth failed to see at Bedford with 5.9 in. [With 3 in, I saw a faint and distant *comes* *s.p*. 27 Jan. 1876].

NEBULÆ AND CLUSTERS.

3041 *h*. 6^h 17^m, *S* 27° 11′. 'Very bright, round, brighter in the middle; 30″ diam. rich' [About 3° North of ζ (3m). A little *n.f* (6^h 18^m, *S* 26° 59′, 1870) is a 'very intense ruby star' of the 8th mag.

3076. (♅ VII. 12) 7^h 2^m, *S* 5° 24′ (1870). 'A fine large rich cluster not much compressed in the middle; stars 9 to 12m' [3° East of γ and about 30′ *f* a 6m star. A good many stars of the larger magnitudes, which seem fainter than 9m. A star about 8½m with a small companion nearly *following*. 3 in. 1875.] Smyth (Cel Cycle) calls this "a solitary yellowish star of the 8th magnitude".

3077. (H VII. 17). 7^h 3^m, S 24°.43′ (1870). Surounding the 5m triple star 30, *n.f* δ Canis Majoris; 'a fine cluster of discrete stars 60 or 70 in number; round; gradually brighter in middle; 8′ diameter. [The component stars which are nearly equal in brilliancy—at least the brighter ones—are well seen with 3 in. aperture].

CAPRICORNUS.

A constellation bounded on the North and East by Aquarius, on the West by Sagittarius, and on the South by Piscis Australis and Microscopium. Its principal stars are α^1, α^2 and β. α^1 and α^2 form a fine pair, visible to some sights without optical aid*. α^1 is double and α^2 multiple. In 1874 I observed the order of brilliancy of the stars in Capricornus to be as follows:— β, δ, α^2, α^1, γ, ζ, ω, ϵ, θ, ψ, η, ι, κ, ν, π, ρ, 24, 36. Heis however rates θ above ω and ϵ and equal to α^1. It will be noticed that the present order of brilliancy does not agree with the order of the Greek letters which

* Heis remarks "Stellæ α^1 et α^2 Capricorni seiunctae a me videntur." I have seen them distinctly in the Punjab sky in full moonlight (with spectacles however, as I am a little short-sighted).

seems to show that probably some variations of light have occurred in the brighter stars of this constellation since the letters were affixed by Bayer in 1603.

Double stars etc.

σ. 20^h. 12^m, *S* 19° 29′ ; 5.5, 10 : 176°.8 : 54″: orange, blue, Webb, who thinks the companion underrated.

[I noticed in Sept. 1874 (3 in.) that the companion was rather brighter than the *comes* to α^1 Capricorni rated by Admiral Smyth at 9½m. In July 1875 also, the companion to σ was much more readily seen than that to α^1].

π. 20^h. 20^m, *S* 18° 38′ ; 6.5, 10 : estimated 145° : 3″.2. Discovered by Burnham [Not examined].

ρ. 20^h. 22^m, *S* 18°. 13′ ; 5½, 8 : 177°.5 : 4″.15 (1836). Webb gives '5, 9 : 3″.8 yellow, ruddy purple', and says 'a 7.5m yellow star *f* considerably *s* seems violet or lilac. Birmingham 2 minute *comites* to 7.5, [Companion small, difficult with 133 on

3 in, perhaps on account of its colour. 7.5m reddish lilac, 1875. One of Birmingham's *comites n.f* readily seen. With 3.9in close pair seen, and 2 *comites* to 7.5 ; the one seen with 3in. quite plain, the other much fainter and more distant]. Sadler with 6½ inch mirror in 1874 noted the colours as yellow and flushed white, and says "only one companion noticed to the 7.5 mag. Distance over 65″ *f*."

ο 20^h. 22^m, *S* 19°. 1′ (1870) . 7, 7½ : 239°.3 : 22″.27 (1836). Webb gives 6, 7 white, bluish 1870'. Smyth (Cel. Cycle) says 'both bluish' [White, bluish, 3 in. 1875. About 1mag, difference. A distant *comes* about 12m. *n.f* nearly in a line with the pair. Some other faint stars in the field]. Sadler with 6½ inch mirror found the colours in 1875 "white, bluish".

Cluster.

3878. M 30. 21^h. 33 ^{m}, *S* 2 °.44′. "Globular, bright, 4′ long, 3′ broad ; all resolved into stars 16m besides a few 12m. Two lines of rather larger stars run out *n. f*". In the same low power field with 41 (5m). [scarcely worth looking at with 3 in aperture. A 8½m star *p*.]

CENTAURUS.

A constellation South of Hydra containing many bright stars, and object of interest. Besides α and β of the 1st mag. there are no less than 9 stars of the 3rd mag. included within its boundaries.

Double stars, etc.

4421. L 4675. 11h. 10m, *S* 47°. 16′ ; 6½, 10 : 68°.7 : 12″. 'Points exactly to a 3rd star'.

4439. B 3594, 3595. 11h. 23m, *S* 42°. 1′ ; 6, 8 : 166°.9 : 12″.93 (1835). 'Fine star.' 8 was strongly suspected to be a close double star by Herschel, who says " if ever I saw a close double star in my life I saw this *as* one" ; this however he failed to 'verify by subsequent examination', and says 'there is no close star. It must have been an allusion, but a very odd one'.

4533. L 5214. $12^h. 30^m$, *S* 39°. 12′ ; 6, 12 : 77°.5 : 30″. In a barren region about 9° North of γ (3m), and closely followed by a 5m star, Lacaille 5231, (4½m).

4539. γ . $12^h, 35^m$, *S* 48°. 18′. 4, 4 : 354°.3 : ¾″ (1836). 'Extremely close and very difficult, 273 barely elongated it'. Webb says 'a rapidly moving' binary. 'Opening, Jacob 1857'.

4500. L 5021. $12^h. 0^m$, *S* 36°. 22′. 6½, 9 : 33° : 50″.

4619. Dunlop 148. 3 (κ) Centauri. $13^h. 45^m$, *S* 32°. 24′. 6, 7½ : 112°.7 : 8″.82 (1836). [White, light blue, 3in 1875. Several faint and distant *comites* in a large field]. It was rated 4½m by Lacaille, and is marked 4m in Proctor's Atlas, and a small 2m (!) in Harding's Atlas. Heis rates it 4-5. [In May 1875, I found it about 5m. In April 1876, it seemed about 4½m. and a little brighter than 1 and 4 Centauri, also rated 4-5 by Heis]. In May 1876, the Revd. S. J. Johnson, with a 4in telescope estimated the components as 5, 6½, and the companion blue.

4624. A 1580. *h* Centauri. 13^h. 46^m, *S* 31°. 21′ : 5, 10 : 185°.6 : 13″.76 (1837) 4in Proctor's Atlas, 5° *n.p* θ (3m). Heis calls *h* 4-5m. [Companion bluish, steadily seen with 83 on 3in. 4 is the most Southern and the brightest of 5 stars forming a curved line].

ξ^2 Centauri. Dunlop 128. 13^h. 0^m, *S* 49°, 16′. 5, 10 : 100°.3 : 6″.01, 1836. [Query, distance 26″.01 ? Herschel elsewhere gives the measurements, 26″ 48 and 25″. 55]

NEBULÆ AND CLUSTERS.

3501. Dunlop 482. 13^h. 18^m, *S* 42°. 24°. 24.′ Sir J. Herschel says "a most wonderful object; a nebula, very bright, very large, a little elongated, of an elliptic figure cut away in the middle by a perfectly definite straight cut 40″ broad; dimensions of the nebula 5′ by 4′. The internal edges have a gleaming light like the moonlight touching the outline in a transparency".

3504. ω $13^h.18^m$, *S* 46°.35′ (1870). Herschel calls this a 'most glorious' and 'truly astonishing' object ; diameter full 20′. It is resolvable into thousands of stars of mags. 12 and 13, and is very visible to the naked eyes as a misty star of about 4½m. The beauty and magnificence of the spectacle afforded by these globular clusters when viewed with powerful instruments is such as cannot be adequately described, and it has been said, that when seen for the first time in a large telescope, few 'can refrain from a shout of rapture.' The component stars, though distinctly visible as points of light, defy all attempts at counting them, and seem literally innumerable. Placed like a mass of glittering diamond dust on the dark background of the heavens, they impress us forcibly with the idea that if each of these points of light is a sun, the thousands, which appear massed together in so small a space, must be in reality either relatively close and individually small, or else the system of suns must be placed at a distance almost approaching the infinite. The former idea is by far the most probable, though it is not easy to imagine, on dynamical principles, how an immense assemblage of bodies filling a globular space can exist in that condition without mutually interfering with each other. At rest they cannot be, as their mutual attractions would soon produce a velocity in each member of the system. They must therefore be in motion, each star perhaps describing its own ellipse about the centre of gravity of the whole mass, which is probably situated near the centre of the sphere.

3555. Dunlop 431. $13^h. 59^m$, *S* 47°.45′ (1870). 'A region of large bright stars, 8, 9 and lower magnitudes ; a very coarse cluster' about 2° *s f.* Centauri.

CETUS.

The largest of the constellations, but not very rich in brilliant objects, the leaders being α and β of the 2.3 mag. β is now distinctly brighter than α, and possibly one (or both) may have changed since Bayer's time. Cetus contains one of the most extraordinary objects in the heavens, Mira (ο), the "wonderful" variable star, further particulars respecting which will be found below.

DOUBLE STARS, ETC.

α and its 5.5^{m} blue neighbour (93 Fl) form a fine open pair, and are, as Mr. Webb remarks, "worth looking at", [Webb's 11m. pair between them, *f*, quite plain with 3in. With 3.9in (Wray), I see two very faint *comites preceding* 5.5 and nearly in a straight line with it].

3420. 37 Ceti. $1^{h}. 8^{m}$, *S* 8°. 34′; 7, 9 : 331°: 60″. Smyth gives 6, 7½, 50″.6. [White, greyish blue, 3in, 1875. A fine pair. Visible to the naked eye between θ and γ, a little North. I found 37 about 5½m, Nov. 1876, and slightly brighter than 44 Ceti].

61 Ceti, $1^{h}. 58^{m}$, *S* 0°, 55′; 7, 11 : 188°·8 : 39″ (Webb). [Companion readily seen with 133 on 3in. Birmingham's 13m *comes* at 55″ glimpsed. The other stars in the vicinity mentioned by Webb also seen. With 3,9in. 11 quite plain, and the 13m *comes* steadily seen].

Smyth (Cel. Cycle) makes no mention of 13. It seems strange how he could have missed it with his excellent $5\frac{9}{10}$ in refractor, as the object is not far South.

3476. $1^h. 52^m. 1^s. 6$. N. P. D 100°. 21′. 9″ (1830). 6, 10 : 183°.7 ; 60″. 'Large star very yellow'. In the place given by Herschel there is no star in Lalande's Catalogue, or Harding's Atlas, the nearest star being Lalande 3620 (6m) *following* ι Ceti.

[Visible to the naked eye *f* ι and nearly in a line with it and *x*. *Comes* small; more like 11.12m. 3in. 1875 (See 5389 in Aquarius). A 7m star, L 3731, *following*, to the North of which is a fainter star, the 3 forming an isosceles triangle. If the object observed by me—which seems to be identical with Lalande 3717 (7m)—is the same as observed by Herschel (as it probably is) the N. P. D as given by Herschel must be 1° in error. The position of Lalande's star, brought up to the same date (1830), is R A. $1^h. 52^m. 0^s. 7$, N. P. D 99°. 21′. 2″, agreeing closely with Herschel if an error of 1° in N. P. D. be admitted].

ι Ceti. $0^h. 13^m$, *S* 9°. 29′. About 2′½ *s. f* this star is a 10m star, and further in the same direction, another About 1° *s. f* Borelly in 1871 discovered a small variable. On Nov. 3, 1871, it was 6½m, but on Nov. 8 it had faded to 8m, and on Nov. 24 to 10m. It remained of this brilliancy till Jan. 1872, since which time it seems to have entirely disappeared (?). A little *n.f* is the variable S Ceti, discovered by Borelly in 1872.

κ Ceti and P I. 182 form a fine open pair, 5, $7\frac{1}{2}$ (Cel. Objects p. 229). Preceding, a little South is a small pair about 11m, and between the pairs a faint star. Heis rated χ as 5-4m. I have seen χ double with a Binocular by Browning.

υ^2 Ceti (4m) $1^h. 54^m$, S 21° 39′ has near it, to the North, a 6m star 57 Ceti. Both stars are of an orange tint in the telescope. A little *s.p* 57 is a 10m star. 57 was rated 5m by Lalande and 6m by Harding, but is not in Heis' Catalogue.

o (Mira) $2^h. 31^m$, S 3° 31′. The wonderful variable star: perhaps the most remarkable variable visible in Northern latitudes. It varies from 2m to a magnitude invisible to the naked eye in a mean period of 331.336 days, or, according to Schmidt, of 330 days, but this varies to some extent. Its brilliancy both at maximum and minimum is also variable. On some occasions at the maximum it equals the 2nd magnitude, and at others does not exceed the 4th (or 5.6, Schmidt 1868). At the maximum of 1779 it was thought to be but little inferior to Aldebaran. According to Schönfeld a minimun occurred on Nov. 20, 1874, when the mag. was 8.6, and a maximum on March 4, 1875 (2.5m).

At the maximum of 1875 I made the following observations on Mira :

24th Jan. 1875. About 8m.

4th Feb. Equal to 63 and 70 Ceti (6m).

8th Feb. Equal to ξ Piscium (4m).

16th Feb. Brighter than δ Ceti (4m) and equal to γ Ceti.

18th Feb. Brighter than α Piscium, but not equal to α Ceti.

25th Feb. Equal to α Ceti, but less than α Arietis

At the maximum of 1876 :

3rd Jan. 1876. About 7m.

12th Jan. About 5½m.

14th Jan. 5m.

17th Jan. About equal ν Piscium (5-4m).

21st Jan. 1876. Very slightly brighter than ξ Piscium.

28th Jan. 4m. and not much inferior to Piscium.

3rd Feb. Equal α Piscium (maximum).

11th Feb. Equal δ Ceti.

11th March. Equal δ Ceti.

On 8th Nov. 1876, about 9m and fiery red.

At the maximum of 1876,—77, I made the following observations :—

5th Dec. 1876. About 5½m, and a little brighter than 67 Ceti.

8th Dec. Brighter than 70 Ceti, but not equal to ρ Ceti (5m. Heis).

11th Dec. Rather less than 5m.

13th Dec. Equal to ρ Ceti.

16th Dec. Slightly brighter than ρ Ceti.

22th Dec. Slightly brighter than δ Ceti (4m and about equal to α Piscium.

27th Dec. About 3m., a little brighter than γ Ceti but more than half a mag. less than α Ceti.

30th Dec. About the same as on the 27th Dec.

2nd Jan. 1877. Rather less than 3m.

5th Jan. Not much brighter than γ Ceti.

To enable the telescopic observer to distinguish Mira—when at or near its minimum—from the small stars in its vicinity, the following alignments may be of use :—About 1° following the double star 66 Ceti (Cel. Objects. p. 229) and a little to the South is a small wide pair. A little following this pair will be found another small pair, and about the same distance further to the East, and a little South, is the small 10m. star which lies about 2′ following Mira (Symth, 88°.9, 116″, 1831). Closely *n. f.* this are 3 small stárs forming a curved line. All these small stars—with the exception of 66 Ceti—are visible in the same field of a low power eyepiece of about 35 or 40 diameters on a 3 in. aperture.

ε Ceti. $2^h. 34^m$, *S* 12°. 22′.was rated once as 3m and twice 4m by Lalande. [In Nov. 1874 and Jan. 1876, I found it about 4½m and equal to ρ]. They are both 5.4 in Heis's Catalogue. vas rated 4.87 by Sir J. Herschel at the Cape.

π Ceti. $2^h. 38^m$, *S* 14°. 22′, was rated 3½, 4, and 4½ by Lalande. It is 4m in Harding's Atlas, and also in Proctor's Atlas and Heis's Catalogue. [In Jan. 1876, I found it very slightly brighter than η Eridani, which Heis rates as 3m. Sir J. Herschel estimated π at 4.56. In Nov. 1876, I found π slightly *less* than η Eridani. It is possibly variable to a small extent]. Ptolemy 3m.

η. $1^h. 3^m$, *S* 10° 48′. This star was rated three times as 3m, and once 3½m by Lalande, 3m by

Harding, 3½ Piazzi, Heis 3.4, Sir J. Herschel 4.06. Smyth (Cel. Cycle) says "it is only marked as of « magnitude by Ptolemy; while Tycho Brahé and Flamsteed make it 3. Had I not adopted Piazzi's magnitudes for my initial star, I should certainly have put this in the 4th degree. Can it be variable?"

[In October 1874, I found it equal to ι Ceti. In Nov. 1876, it seemed slightly brighter than ι Ceti; also very slightly brighter than θ Ceti, which Heis rates at 3m].

Star in Harding's Atlas.

R A. $1^h. 50^m$, *S* 20°. 56′.

A little *n.p* ν^2 Ceti (59 Flamsteed). Marked 6m by Harding, but not in Lalande's or Heis's Catalogues. About 13′ *s. p* is the star Lalande 3590 (8½m.). [Nov. 18. 1876, I found Harding's 6m star about 8m, being then brighter than L 3590 but considerably less than a 7m star (Harding) closely *p* ν^2. Harding's star may be variable and should be watched. About 1° *s. f* ν^2, Harding shows an 8m star, which is identical with Lalande 3798 (8m). This is now about 6½m. and slightly brighter than the 7m star *p* ν^2.].

COLUMBA.

A small constellation, South of Lepus. Its principal stars are α and β of the 3rd magnitude.

Double stars.

3728. B. 881, 5^h. 5^m, *S* 41°. 23′; 6½, 11 : 262°: 14″. 'Fine'. *South preceding* the nebula 508 Dunlop.

3732. L 1753. 5^h. 8^m. *S* 27°. 19′ ; 8,8 : 216°.5: 80″. The most Southern of 4 stars forming a crooked line (in Harding's Atlas). [Both light orange, 3in, 1875. A fine open pair].

3735 . L 1759. 5^h. 9^m. *S* 32°. 2′ ; 8,8 : 150°.8 : 7″.59. 'Fine double star'. [About 3.° *n. p* ο (5m), the most Northern of two stars of equal brilliancy].

3740. L 1780. 5^h. 11^m, *S* 36°. 47′ ; 7″,8½ :

284°.7 30″ ; yellow, blue [a little South of ο (5m) 8½ lilac, 3in, 1875].

3760. 5ʰ. 22ᵐ. *S* 35°. 27′ ; 7, 7½, 11 : 221°.3, 270° : 7″.32, 20″. [Between ε and ο ; 11m. glimpsed with 133 on 3in, 1875].

3849. 6ʰ. 16ᵐ, *S* 39° 26′; 6.5,7 : 52°.3 : 39″.43. 'Large star yellow'. [A little W of a line joining κ Columbæ (4m) with ν Puppis (3m) and nearly in a line with γ and θ. Both stars the same colour, 3 inches 1875. Two faint and distant comites, *s.p* and *s.f*].

Clusters.

2730. 5ʰ. 1ᵐ, *S* 38° 8′. " Very bright ; very large (5′ long by 2′ broad) ; much elongated ; stars seen in it. Visible with moon and lamp illumination ".

About 3° *n.p* 2777 *h*.

2777 *h*. 508 Dunlop. $5^h. 9^m$, *S* 40° 11′. "Superb globular cluster ; very bright ; round, 4′ ; central blaze ; stars 14m ; the stars barely visible in strong twilight". About 2° *n.f* the double star 3728 *h* (above).

CORONA AUSTRALIS.

A small constellation South of Sagittarius, and close to the tail of Scorpio. Its principal stars which are of the 5th and 6th magnitudes are arranged in a curved line, forming a good example of a "star stream" . κ (6m) was suspected by Halley to be variable (3m to 6m). There are two small variables between γ and ε . In Lacaille's Catalogue α is rated 4½m and β 4m.

Double Stars etc.

5046. Dunlop 222. κ. $18^h. 25^m$, *S* 38°. 49′ : 6,7 : 359°.3, 21″. 78 (1836). [White, reddish yellow, 3in 1875. A faint and distant *comes n. f.* A few minutes to the North is an 8m star]. κ is a suspected variable.

5093. 19_h. 4^m, *S* 43°. 20′. Triple 8,10,11. A B. 220°.2 : 20″. B C. 213°.6 : 10″.

5084. γ 18^h. 57^m. 48^s. *S* 37°.14′. 18″ (1875) 6,6 : 37°.1 (1834) ; 36°.8 (1835), 34°.5 (1836), 32.7 (1837). 2″.66 (1837). Herschel says there is no "shadow of doubt as to the reality of a considerable retrograde orbitual motion of 1°.47 in this beautiful double star". It was measured by Capt. Jacob in 1858. 343°, 1″.53, and he assigns a period of about 101 years ; the periastron passage occurring in 1863, when it was measured by Powell, 318°.1 : 1″.25. It is now widening again, and is probably within the reach of a 3 inch aperture. [I failed to split it with 3in 1875, but it seemed elongated in the direction P=250°±].

CORVUS.

Is a small constellation South of Virgo, and North of Hydra. Its four principal stars, β, γ, δ and ε form a trapezium. α is possibly variable, being now only of the 4th magnitude. If as faint in 1603 as at present, it is difficult to understand why Bayer should have made it the principal star in the constellation. ζ may also be variable; it is marked 3m. by Harding and was rated 5, 5½m. by Lalande and 5m. by Heis. β has also been suspected of variation (See Cel. Objects p. 234). Heis rated γ Corvi as 2m, but it is certainly less than the brighter stars in the Plough (2m. Heis). In March 1876, I found γ about equal to κ Orionis (3.2 Heis).

DOUBLE STARS.

δ 12^h. 24^m, *S* 15°. 50′. 3½, 10 : 213° : 23″.44 (1837). [The magnitude of the small star seems underrated by Herschel. Webb gives "3, 8.5, light yellow, purple". Heis rated δ at 2.3m. I saw the *comes* with reduced apertures of 2in. and 1½in. (of 3 inch refractor) in 1875. δ seems to me nearer 3m. than 2m. Sir J. Herschel estimated it 3. 22]

4505 12^h. 6^m, *S* 16°. 7′; 6, 8 : 280°: 8″. A little *n.p.* γ Corvi (3m). [Companion small with 3 in 1875 ; more like 9m. The primary is also less than 6m. It was rated 7m. by Lalande and Harding. There is a 10m star, a few minutes to the north, a little *f.* I glimpsed a closer and fainter *comes* to the South; the three stars (4505 and the 2 *comites*) nearly in a straight line].

CRATER.

A constellation adjoining Corvus, to the West, and North of Hydra. Its principal stars are of the 4th magnitude.

Double stars etc.

4502. $12^h.3^m$, S 11°. 10′. Triple, 7, 9, 8 : 90°, 93° : 12″, 60″. Herschel says "A and C white, B very ruddy. About 5° *North* of, and a little *preceding* γ Corvi.

Red star. R Crateris. 11046, O. A. $10^h.55^m$, *S* 17°. 40′. Closely following α (4m); "intense scarlet star, a curious colour". Baxendell found it to be variable [Full scarlet, 3in, 1875. One of an open pair, and the further of the two from α. About 8½m. (1875) and brighter than the star preceding. There is a third star about 9m, a little South of it. Ward sees a 13m *comes* between α and R, with a 2¾ inch Wray, [13m. readly seen with 3 in. March. 1876]. Smyth (Cel. Cycle) makes no mention of this faint star, although he observed the three others. With reference to α Crateris, Smyth remarks ; "This object may once have been brighter, since it acquired a name—Alkes—and was lettered α ; but δ is now the lucida, and wears the Greenwich honours". α was rated 4m. by Heis and δ 3.4. [In March 1875 I found δ slightly less than ν Hydrae also rated 3.4. by Heis].

ERIDANUS.

A large constellation which extends from Orion to nearly 60° South Declination.

Double stars etc.

3493 . φ . 2^h. 12^m, *S* 52°4′. 5, 9 : 210°.4 : 80″. φ was rated 3½m. by Lacaille in 1751, and 4m. by Sir J. Herschel at the Cape. [I found it brighter than κ and χ Eridani, and about 4m. Dec. 1875].

3533. 2^h. 46^m, *S* 20°44′. 8, 8½ : 274°.7 : 45″, 'yellowish, bluish'. North of τ^2. [A fine open pair, reddish, bluish, 3 inches 1875 . A little South of τ^1 are 3 small stars nearly in a straight line, and about midway between two smaller stars].

3545. θ. 2^h. 54^m, *S* 40°. 47′. 5,6 : 81°.5: 8″.68. "Superb double star". Webb says 'moving, Jacob

1857'. [White, light yellow, 3 inches 1875; light yellow, dusky yellow, 1877. A splendid double star; one of the finest in the heavens. The magnitudes seem to me much underrated; more like 4, and 5, or 3½ and 4½. Sir J. Herschel at the Cape rated θ at 3.73, and Lacaille gave it 3m. I found it about 3m. Dec. 1876, and considerably brighter than 12 Eridani. If 'moving', the motion must be very slow, as I estimated the angle at 80°± Jan 1877].

3556. 12 Eridani. 3^h. 7^m, *S* 29°30′; 3½, 8: 310°: 5″.3 (1836). Herschel says 'Some suspicion of elongation in the large star'. Binary. Jacob in 1856 found it 'closing rapidly'. [Companion not seen with 133 on 3 inches 1875].

3597. *f* Eridani. 3^h. 44^m, *S* 38.° 0′. 5, 5½; 199. 7: 8″.55; Superb double star'. Suspected to be a binary by Jacob in 1856. South of υ^2 (3½m) [I found the components yellowish white and very light green, 3 in, 1875. In Jan. 1875 I noticed that *f* was about equal to υ^2. They were both rated 4m. by Lacaille in 1751].

3608 γ Eridani. $3^h. 55^m$, *S* 13°. 42′. 3½, 13 : 223°. 6 : 45″. Herschel says "marked as 2.3m. in *A. S. C.* but it is certainly not above 3.4". It was rated 2m. by Lacaille and is marked 2m. in Harding's Atlas (1822). In Herschel's estimates of magnitudes it is 3.94. Heis rates it 3m. [In Nov. 1875 it seemed to me about 3½m. 13 glimpsed with 3.9 inches]. The 13m. companion is not given by Smyth, but he notes a distant 10m. star *n. p.* and an 11m. *s.p.* [In Dec. 1876. γ seemed about 3m. and equal to β Eridani]. D'Arrest calls γ 'orange'.

3621. 3^h. 59m, *S* 34°. 10′; 8½, 8½ : 110°.3: 15″. 'Fine star'. Between υ^3 and υ^4 (not examined).

3628. $4^h. 9^m$, *S* 36°. 36′; 7, 7: 50°: 50″.51 (1838) *s.p* υ^4. [A faint *comes* at a considerable distance, 3 in, 1876. A very faint pair *follows*, a little *South*].

3628. Struve 316. 39 Eridani. $4^h. 9^m$, S 10°. 33′; 6, 11 : 140°± : 5″. Herschel also gives 6½, 12 : 146°.6. and says 'Query if not changed in position'. Webb gives '5, 11. (perfectly easy) 154°: 7″.1

(full yellow, deep blue'. It is called A in Lálande's Catalogue, and rated 5m, and also 6m. Heis rates it 5m. It lies a little South of o^1 and o^2 [Companion readily seen with 3 inches, Nov. 1875. A beautiful object. Nearly points to a small star *s. f.*]. Smyth (Cel Cycle) calls it a "delicate double star," but it is not a difficult abject with 3 inches.

3636. υ^4. 4^h. 13^m, *S* 34°.5′. 3, 13 (or 14): 14°.6 : *55″* [*Comes* not seen with 3 inches. In June 1876, I found υ^4 brighter than υ^5, and about equal to γ or 3½m.].

τ^4 (16 Fl) 3^h. 14^m, *S* 22°.11. 3½, 11: 240°. 8 : 150" "light orange, greyish"—Smyth—who says, (Cel Cycle) "this object is in a barren field, and the large star seems overrated, since it appeared more than once diminished to nearly a 5th magnitude; but the lowness of its position renders the case doubtful, from variable refraction". It was rated 4m. by Lalande, and 4.3 by Heis. [In Jan. 1876, I found it about equal to τ^3 and slightly brighter than τ^5 (4m Heis). In Dec. 1876, it seemed a little brighter than τ^3, and of an orange hue.

Symth's 11m. companion seen with 3 inches; very faint ; more like 12m.]

[About 3° South of ι Eridani is a star marked 6m. in Proctor's Atlas. In Nov. 1874, and Dec. 1875., I found this star very nearly equal to ι. It is called *s* Eridani in Lacaille's. Catalogue, and rated 4m. or equal to ι.]

64 Eridani. $4^h. 54^m$, *S* 12°.43′ was rated 6m by Heis or equal to 63, a little North of it.

[In March 1876 I observed 64 much brighter than 63 and nearly equal to κ Leporis (5m. Heis). In Dec. 1876, I found 64 slightly brighter than ν Leporis (5.6 Heis). It may possibly be variable and should be watched].

NEBULA.

2521 Dunlop 487. $3^h. 13^m$, *S* 41°. 32′. "Globular; very bright ; first gradually, then suddenly very much brighter in the middle; resolvable, mottled, but not resolved."

FORNAX.

A small constellation, South of Cetus and Eridanus. It contains no star brighter than the 5th magnitude, but there are some double stars worthy of notice.

Double stars.

3484 . 2^h 4^m, *S* 30°. 13′. 8, 9½: 63°.5: 89.″16. A little North of a line joining μ and ν (5m.). Not in Harding's Atlas.

3506, ω. 2^h. 29^m, *S* 28°. 45′; 5½. 8: 241°.7: 11″.15. Herschel says 'very fine star, but ill defined'. [Yellowish, lilac, 8 more like 9m, 3 inches, 1875. Two faint and distant *comites South preceding* form with the pair an isosceles triangle].

3532. υ. 2^h. 43^m, *S* 37°. 54′. 6½, 8 (or 6, 7) 150°.6 : 8″. North preceding θ Eridani [yellowish, bluish, 3 inches, 1875. Companion small, more like 8 or 8½ than 7m. I found υ about 7½m. and much less than η^2 and η^3 Fornaci. Dec. 1876].

γ′ Fornacis. $2^h. 45^m$, *S* 25°. 3′. 6, 12 : 171°: 45″, (1837) Smyth (Cel. Cycle) who says, "A wide but delicate double star". The *comes* should however be within the reach of a 3 inch aperture. [I found γ′ a little fainter than ζ Fornacis in Dec. 1876. Smyth's 12m. *comes* readily seen with 133 on 3 in. Jan. 1877].

GRUS

Is a constellation South of Piscis Australis. Though small it contains some bright stars, and β being of the 2nd magnitude. The leading stars, with the exception of α, form a curved line.

Double stars.

5295. B 7080 . $21^h. 40^m$, *S* 47°.51′. 6½,9½ : 14°. 2: 30″.3. Herschel says with reference to measurements of this star "all circumstances favourable". [About 5° preceding α (2m). Small star bluish. Well seen with 40 on 3 inches, though rather low in the Punjab sky].

5338. Dunlop 239. δ^2. $22^h. 23^m$, *S* 44°. 22′.

5, 9½: 215°.7 : 61″.26 (1836): 'yellow, bluish' [Small star blue, 3 inches, 1875. South, a little *f* is a 7m star, Lacaille 9149. δ^1 and δ^2 form a fine pair].

5383. Dunlop 246. -8046. B.A.C. 23^h. 0^m, *S* 51°.23′ (1870) 6½, 7: 260°.5 : 8″.12 (1836). 'Very fine star', 'changed (?).

Nearly in a line with μ and β.

HYDRA.

A long straggling constellation, extending nearly East and West, and South of Cancer, Sextans Crater, Corvus, Virgo and Libra. It is particularly devoid of brilliant stars, the only bright one being α, a reddish star of the 2nd magnitude which was thought by Sir J. Herschel to be slightly variable. None of the other stars South of the Equator are above the 4th magnitude except γ and ν which are small 3m. stars.

It contains however some interesting doubles.

Double stars etc.

α. 9^h. 22^m, *S* 8°, 8′. Though to the naked

eye deserving the name given to it by the ancients of Alphard or the "solitary one", "it is by no means a completely isolated star when examined with the telescope. It has a faint and distant *comes* observed by Smyth, and about 25′ *p*, Ward sees a small double star, estimated 8, 13: 90°: 50″± [With 3in. I see a small star about 9½m, a few minutes to the South and a little *f*, probably identical with Smyth's *comes*. Further in the same direction is a fainter star, and others at greater distances in the field. There is also a faint star a little to the North. Ward's double seen with 133 on 3 inches]

4149. 8^h. 50^m, *S* 17°. 47′. 5½, 7 : 145°.5 : 68″.95. [Barely visible to the naked eye, even in the Punjab sky; the magnitndes are apparently overrated. It is not in Heis's Catalogue. In Lalande's Catalogue (17637—38) the magnitudes are rated 6½, 8, which is more correct. A fine open pair; both orange, 3 inches 1875. A few minutes North is a fainter and closer pair about 11m.].

κ Hydrae. 9^h. 34^m, *S* 13°. 47′. 5m in Proctor's Atlas, 4m Harding, 4m and 6m, Lalande. Heis rates it 5m. [I found it about 5½m, 25th April 1875; not quite equal to ν′].

4446. 17 Hydrae (17 Crateris in Cel. Objects). $11^h.26^m$, $S\,28°.32'$ (1870); 6, 6; 30°.2: 8″.92 (1835). The *s.p* star is probably variable as Webb gives "5.5,7 : 207°.8 : 10″ (1833); (211°.4; 8″.8. 1857), lucid white, violet," and says '7 var ?, only ½m. difference 1852. So Wrottesly 1857' [About 2°½ North of ξ (4½). A fine pair; well seen with 40 on 3 inches, 1875: both yellowish; the *s.p* star rather the larger. There is a 7m. star *north following.*]

4449. ξ. $11^h.27^m$, $S\,31°11'$; 4½, 11: 170°±: 60″ [11 glimpsed with 3 inches, 1875 ; more like 12m. There are several other small stars in the field. (See note to 4133 in Argo)]. Lacaille rated ξ as 3½m. Heis gives it 4m. In May 1876, I found ξ about ¼ mag. brighter than β Hydrae, also rated 4m by Heis.

4463. $11^h.35^m$, $S\,32°\,55'$; 6, 8; 248°. 3: 2″½; 'large star yellow'. A little South of a line joining ξ and β (4m) and equidistant from those stars. [Not seen].

4478. β. 11^h. 47^m, *S* 33°. 14′. 5, 5: 340°: 2″ (1835). Herschel says a 'beautiful object, very like ξ Ursae, but closer, and a little more unequal'. A suspected binary, Herschel states, 'beyond a doubt,' but Webb in Celestial Objects says 'not confirmed by Jacob'. [I failed to divide this star with 133 on 3 in, April 1876.]

4554. L. 5301. 12^h. 45^m, *S*. 30°. 26′; 6, 10: 28° $\pm$: 18″. Lacaille rates it 7m. It is the more Northern of two stars close together. [*Comes* small with 3 in. 1876. A more distant and brighter companion nearly due North. The large star is red and about a mag. fainter than the 7m. star to the South of it].

4597. 138 Dunlop. *f*. 13^h. 31^m, *S* 25°. 53′; 6,7: 191°.5: 10″.04 (1834). 'Fine' [About 5°. *s.f* the red star γ Hydrae (3½m). Yellowish, bluish, 3 in. 1875].

R (ν) Hydrae. Variable. $13^h. 23^m$, S 22°. 36′. 4m to below 10m in about 437 days. Discovered by J. P. Maraldi in 1704. It is also red. At the maximum its brilliancy varies from 4m to 5½m Its period according to Schönfeld has diminished since the date of its discovery, when it was about 500 days. According to the same eminent authority, a maximum occured on 25th Feb. 1875, and the next (predicted) on the 10th May 1876. From Schmidt's observations, the minimum seems to happen 200 days before the maximum. Heis gives the minimum as 11m, and the period 448 days. Smyth (Cel. Cycle) calls this star μ Hydrae, but this is a mistake, the real μ (42 Flamsteed) being a 4m star, a little South of a line joining λ Hydrae with Crateris. [In 1875, I made the following observations on this remarkable variable:—

5th Jan. 1875. Not visible to naked eye.

10th Feb. About equal to ψ Hydrae (5m Heis). (max).

3rd March. Not quite equal to ψ

30th March. Decidedly fainter than ψ

25th April. Very small in opera glass or about 7½m.

And in 1876, the following observations:—

30th March 1876. A little fainter than ψ Hydrae.

12th April. About equal to ψ (maximum).

10th May. Slightly less than ψ.

19th May. Rather more than ½ a mag. less than ♆.
23rd May. About 1 magnitude less than ♆.
11th June. About 7m.]

15 Hydrae. $8^h. 46^m$, *S* 6°. 25′. Triple, 6½, 12, 13 : 345°, 43°, 40″, 55″. About 12° West of α Hydrae.

[Smyth calls this "a most delicate triple star", but I saw both companions distinctly with 133 on 3 in, April 1876].

4698. 54 Hydrae. $14^h. 39^m$, *S* 24°. 57′; 6, 8½ : 134°.2 : 9″.19. Numbered 10 in Harding's Atlas, and the most Northern of a curved line of 6m stars preceding 20 Librae (3½m). Smyth (Cel. Cycle) also calls it 10 Hydrae, and rates the components as 5½, 7½.

Red Star $13^h. 42^m. 15^s$, *S* 27°. 46′ (1880). Discovered by Burnham in April 1876. He describes it as "one of the finest object of the class". Dembowski called it "a perfect blood red" and rates the magnitude as 6.8. Burnham adds that it "has a blue 10m.. companion at a distance of about 77″ in the direction of 110°.2, and identifies it with Lacaille 5686.* It lies *s.p* π Hydrae, and about 4° *n* of 3 (κ) Centauri.

LEPUS.

A small constellation, South of Orion. It contains no star brighter than the 3rd magnitude. μ is shown in most star Atlases as 5m, but it is certainly brighter than λ, rated 4½m. Heis correctly rates μ as 4m. There are many discrepancies of this sort to be found in star maps.

Double stars, etc.

3745. 5^h. 14^m, *S* 18° 38′. 7, 7½: 18°: 39″ (1837). 'Superb star', a third star C, 2′ ± distant is situated a very little following the line A.B prolonged to C—9m'. [So I see it. Between β and μ. Both white and almost exactly equal. A distant and very faint *comes* nearly following glimpsed with 3 inches, 1875. Preceding 9 is a faint star, and following is a 9½m. star, with a small companion to the South].

* Astronomical Register No. 164. August 1876.

3752. ε. B.A.C. 1670. $5^h. 17^m$, *S* 24°. 54′. Triple 6, 9½, 9: 110°.3, 106°.1 : 3″.33, 58″.81 (1837). [Visible to the naked eye, about 4° South of β and in a line with it and α. 9½ seen with 133 on 3 in. An 11m star *n.f.* In the field of a 40 power eyepiece, *n.f* is the cluster 79 Messier, a bright object, between two 8m stars]. Under 79 M. Smyth (Cel. Cycle) mentions ε Leporis as a "fine white star with a red companion of the 7th magnitude in the *n.p* quadrant."

3761. β . $5^h. 23^m$, *S* 20°.51′. 4, 13 : 146°: 70″. [Comes not seen with 3 in. Readily seen with 3. 9 in. Seems less than 13m. β was rated 3.4 by Heis (α being 3m). From estmations of its magmitude made by Sir J. Herschel at the Cape he considered β to be "probably variable". In Dec. 1875, and March 1876, I found β to be almost exactly equal to α. Both about 3m]. The 13m *comes* is not given by Smyth (Cel. Cycle) but he notes an 11m companion distant 210″. P 67°.5 [Seen, 3 inches]. The primary star was discovered by Burnham (1873) to be a close double, the companion being 11m, and distant 2″.89. P. 267°.1. He compares it to Antares.

3780. $5^h.34^m$, *S* 17°. 59′. Multiple. Closely *following* α. Herschel merely gives the magnitudes as 7, 7, 8, 8, 8 without positions or distances. According to the eminent American observer, Burnham, the details of this object are as follows:—Mags. 7, 8½, 9, 8, 12, 11 (9, 10): P, 6°.7, 136°.2, 298.7, 45°.8, 104°.5, 7°.6 : D. 78″, 90″, 125″, 50″, 90″ (1″.5.). The close pair was discovered by Burnham, who also found the principal star, 7, to be a very close and difficult double. [The 6 principal stars well seen with 3 inches. The close companions are quite beyond the power of a 3 in aperture. A beautiful object for a small telescope, 11, 12 and 8½ lie nearly in a straight line].

γ. $5^h.39^m$, *S* 22°.29′; 4, 6.5, 13 : 349°, 345° : 93″, 138″; pale yellow, garnet, Webb 1851. [I found 13 quite plain with 133 on 3 inches, 1875. The three stars are nearly in a straight line. I glimpsed a fainter star further to the North]. Smyth (Cel. Cycle) notes a distant 12m *comes following*.

LIBRA.

One of the constellations of the Zodiac, preceding Scorpio. Its brightest stars are α and β of the 3rd magnitude. The star δ, *preceding* β, is variable from 4.9 to 6m. in 7.3 days, discovered by Schmidt. Heis however gives the period as 2d. $7^h. 51^m. 19^s$.

In Proctor's Atlas ζ^1 is marked 4m and ζ^3 and ζ^4 6m. Harding shows ζ^4 the brightest of the three, and so I found it in May 1875.

ι, *South following* β, was rated 4m by Lalande and Harding, and 5m by Heis. In May 1875, I found it rather fainter than 37 (5m Heis) or about mag. 5.6. It may possibly be variable.

DOUBLE STARS.

β. $15^h. 11^m$, *S* 8°, 56′, is said to be a pale green hue to the naked eye. It is a wide double star, or rather has a very distant companion, according to Smyth 12m and distant 570″ on an angle of 85°.3.

α^2, α^1 $14^h. 44^m$, *S* 15°. 32′; 3,6 : 314°.3 : 229″ pale yellow, light grey (Webb). [A fine pair]. The position above given is that of α^2, the brighter star.

4719. A 1690. $14^h. 50^m$, *S* 20°. 47′; 5½, 7 :

277°.4 : 12″.08 (1837) "large star yellow, small more yellow ; a curious coloured star, not a bad miniature of α Centauri". It is Lalande 27173 [A faint star *p* a little *n*. A brighter and more distant one to the North]. Smyth (Cel. Cycle) mentions another companion 16m, pale red, distance 20″. P 320°, discovered by Herschel, and adds "the comes is the very *minimum visibile* of my refractor"

4783. 15^h. 24^m, *S* 19°.44′; 6, 10 : 281°.7 : 11″.06 (1836). 'Neat star'. Smyth (Cel. Cycle) gives the components as "A 7½ bluish white ; B 9 smalt blue, with two or three minute stars in the field."

LUPUS.

A constellation South of Libra and Scorpio, and interesting to the amateur observer from the fact that so many of its leading stars are doubles, and easily found without an equatorial mounting.

Double stars.

4739. ζ Lupi. 15^h. 3^m, *S* 51°. 36′. 4½, 6 (or 7): 247°.6: 80″.

4739. κ. $15^h. 4^m$, *S* 48°.17′; 5, 6½: 144°.2: 27″.21.

4752, 4753. μ Lupi. $15^h. 10^m$, *S* 47°. 26′. Triple, 6, 6, 8: 173°. 6, 131°.3: 2″.08, 22″.93 (1836.) .About 3°½ North of ζ (4m). Rated by Lacaille as 4m in 1752.

4760. ε $15^h. 15^m$, *S* 44°.16′; 4, 9½: 174°.9: 26″.35.

4819. Dunlop 196. ξ. $15^h. 49^m$, *S* 33°. 36′; 6, 6½: 49°.3: 11″ (1835.) ‘Superb double star’. The *following* of two 4m stars in Proctor's Atlas near each other. [Both light yellow, 3 in, 1875. A fine object for a small telescope. ξ fainter than χ. Heis rates it 5m].

4823. η Lupi. $15^h. 51^m$, *S* 38°.2′; 4, 8½: 22°.2 15″.07 (1834). ‘Small star blue’. [Small star seemed lilac to me with 3 inches, 1875. Two small stars *precede*, of which the Southern is the brighter].

183 (?) Dunlop, Lacaille 6361. $15^h. 18^m$,

S 38°. 19′; 5, 9½, 9½ : 298°.7 : 45°.9 : 89″.48, 144″.5. About 1°. *n* of *υ* (6m).

4831. B 5598. 15^h. 59^m, *S* 36°. 29′; 6, 12: 340°: 40″. 'The middle of 3'. Herschel also gives the magnitudes as 6, 13 and 5½, 11½. It lies closely *north following θ* (5m). [*Comes* not seen with 3 inches 1876].

MICROSCOPIUM.

A small constellation South of Capricornus and between Sagittarius and Piscis Australis. Its only conspicuous star is *α* of magnitude 4½.

Double stars.

5224. *α*. 20^h.42^m, *S* 34°. 9′; 5, 11: 164°.5: 21″.71 (1835) [Comes difficult with 3 inches, 1875].

5238. Dunlop 236. 20^h. 53^m, *S* 43°. 28′; 7½, 7½: 254°.9: 60″ (1836), South preceding *η* (5m). [A fine open pair, both yellowish, but some difference in colour, 3 in. 1875. The *s. p* star rather the larger of the two].

OPHIUCUS.

A large constellation, South of Hercules and North of Scorpio. The only bright stars it contains South of the equator are δ, ϵ, ζ and η, all of the 3rd magnitude. Between δ and ϵ, Harding and Proctor show a 6m. star. I noticed in Aug. 1874 that this was quite invisible to the naked eye. Ophiucus is remarkable as being the constellation in which two Temporary stars have appeared and afterwards vanished. One seen by Kepler in 1604 was of great brilliancy and further particulars respecting it are given below. The second was noticed by Mr. Hind in 1848 between η and the 5m star 20.

Double stars etc.

4854. 5 or ρ Ophiucii. $16^h. 18^m$, *S* 23°. 10′, Quadruple, 6, 6, 8, 8: 2°, 2°, 286°.6. Distance of close 6m stars 4″.78. Webb gives '5, 7.5: 3°.1: 3″.8 and the colours as 'pale topaz, blue'. Herschel says "No other star but its 2 companions on an intensely black ground" [ρ in Proctor's Atlas, North of Antares. With 3.9 inch, I observed a faint *comes* about 13m to the South, and more than double as far from the primary star as either of the 8m. conpanions. June 1875. If Herschel did not see this small star, it must be variable, as it is now plain with 3.9 inches. I glimpsed it with the 3 inch, July 1875. In May 1876, I found it nearly steady with 3 inches].

4901. $16^h. 50^m$, *S* 19°. 22′; 6½, 8 : 229°.6 : 5″. This seems to be identical with the star marked 236 Piazzi in Proctor's Atlas *s. p.* 29. Ophiucii.

4928. 36 Ophiucii. $17^h. 8^m$, *S* 26°. 25′; 6, 6: 43°.5: 4″.97 (1835); ruddy, pale yellow. Webb gives from Smyth 4.5, 6.5: 226°.1: 5″.2, 1831. The *f* star is probably variable which would account for the difference in the position angles. Webb says '1854 nearly equal, about 6. 5, Smyth's smaller perhapes rather the larger'. Herschel says 'as good measures as it is possible to get. This star has changed' (this remark however probably refers to the position angle). Binary. Period about 200 years. Closing [Both yellow, and almost exactly equal. The *f* star, if anything, rather the brighter of the two. In the field *n. p.* is a 7½ or 8m star, to which I see a faint *comes* to the South, 3 in 1875]. Smyth says "The principal star is thought to be variable, though I have always seen it as now registered" ($4\frac{1}{2}$, $6\frac{1}{2}$), and he calls the companion to $7\frac{1}{2}$ "a most minute *comes*" 'whose existence, Sir John Herschel first pointed out' to him. In a large field with 36 Ophiucii is the star 30 Scorpii. Between the two, Smyth (Cel. Cycle) notes a 14m star. [Close to the place indicated by Smyth (in a diagram) I see steadly with 3in. a small star; other faint stars in the field].

4937. 5827. B. A. C. 39 Ophiucii. 17^h. 10^m, S $24°$. $9'$. $5\frac{1}{2}$, 6: $358°.3$; $15''$ 'Superb double star'. Smyth gives $5\frac{1}{2}$, $7\frac{1}{2}$ [Yellowish, bluish; about 1 mag. difference 3 in. 1875].

Nova 1604. 17^h. 23^m, 9^s, $S21°$. $22'$. $16''$. (1875). About midway between ξ (5m.) and 58 (5m.) Ophiucii. Kepler's Great New Star which blazed out in Oct. 1604. When first seen it was, according to Arago white and exceeded in brilliancy Mars Jupiter and was even thought to rival Venus in splendour! It gradually diminished however and in 6 months was not equal in lustre to Saturn; in March 1606 it had entirely disappeared. The position assigned above to this wonderful object is that deduced by Prof. Schönfeld from the observations of David

Fabricius. There does not however seem to be the same amount of certainty with reference to its exact position as in the case of Tycho Brahé's star in Cassiopea, nor is there any star within 2′½ (of arc) of the place determined by Schönfeld, the nearest star being one of the 12th magnitude, a little *s.f.* (See Frontispiece). Chacornac however in 1861 mapped a star of 10th mag. about 2′ *preceding* the spot; this star would seem to have since disappeared, as it was missed by some observers in 1871 and 1872* Prof. Winnecke however in 1875 observed a 12m. star very near the place occupied by Chacornac's star, so that a variable star would seem to exist close to the place of the Nova, and its variability is of course evidence in favour of its identity with the long lost star of Kepler. The spot should be carefully watched with a large instrument. About 6′ *following* the place is a 9m. star, Oeltzer 16872. [This star is again followed by two fainter stars about 10½m. forming with it an isosceles tringle. The faint stars near the place of the Nova not seen with 3 inches].

Nova 1848. R A. 16h. 52m, 13s. *S* 12°. 41′.24″. Discovered by Hind, 28. April 1848. When first

* See *Nature* Jan. 28, 1875, and Aug. 12, 1875.

noticed it was about 5m.; it afterwards rose to 4m, but very soon faded to 10 or 11m. Its place is about 18′ *north* of, and a little *following*, a 9m star Lalande 30853, which again is about 30′ *n. p.* a 7m star L 30903. These two latter stars are nearly on a line joining η (3m.) with 20 (5m). There are several small stars near its place.

This curious object has become very faint of late years. In 1856 it was 10m; in 1866, 12m; and in 1874 and 1875 not above 13m.

Star. R A. $17^h. 8^m, 29^s$, *S* 14°.27′. This star which lies between η Ophiucii (3m) and ν Serpentis (4½m) was discovered to be double by Burnham (No. 250 of his Fifth Catalogue of New Double Stars) who suspects it to be variable, as it is not found in the Catalogues of Weisse, Lalande Rümker and others †. It was estimated 7m. by Burnham and twice 6.5 and once 7m. by Dembowski. It is 6m. in Harding's Atlas, but is not given by Heis. The small companion (distance 4″.2) is very faint, being, according to Dembowski, only 12m of Struve's Scale. Burnham Considers it a "pretty good test for a 6 in. aperture."

---

* *Nature* June. 29. 1876.

† Astronomical Register. No. 164. August. 1876.

[In Sept. 1876, I found the star 6½m and about equal to a small star (7m Harding) closely *f* η Ophiucii. Not equal to a star about 3°. *s. p.* η Ophiucii (rated 6.7 by Heis, 6m Harding.) Burnham's star *just* visible to the naked eye of a clear night in the Punjab, Sept. 8. 1876].

ORION.

This magnificent asterism—the finest in either hemisphere—is full of interesting telescopic objects, many of which can be well seen with a small telescope in Southern latitudes.

δ is said to be variable to a slight extent, from a maximum—according to Heis—of 2.2m, to a minimum of 2.7m.

Double Stars.

β. Rigel. $5^h. 9^m$, *S* 8°. 20′; 1, 8: 199°.8 : 9″. 42 (Sir J. Herschel). Webb gives the magnitude of the companion as 9m. This fine object is a well known test for small apertures. It may be well seen with a really good 3 inch achromatic, though many sold as such will not show a trace of the companion. [I have seen it well in the Punjab sky with the 3 inch refractor reduced by a stop to 2 inches.

With 3.9 in. the companion was quite a conspicuous object. It may be considered a good test for a 2¼ inch aperture of high quality].

ζ. $5^h. 35^m$, *S* 2°.0′; 2, 6 : 147°.9 : 2″.76 (Herschel). A 10m star, 7°.8, 56″—Webb. who says the close companion "seems of some nondescript hue, about which observers cannot agree". [I found it green with 3 in, 1875]. Some observers suppose the 10m *comes* to have brightened of late years, and this seems probable, as the Revd. S. J. Johnson has seen it well in England with an aperture of only 2¼ inches. He considers it to be not less than 8m, but judging from its visibility in a 3 in. refractor, I consider it to be certainly *below* 9*m*. In Jan. 1875, however, I saw it steadily with a 2 inch stop, on 3 in.

ι 5^h. 29′, *S* 6°. 0′ (1870); 3½, 8½, 11 : 141°.7, 102°.8 : 11″.5, 49″ (Webb). The 11m. *comes* may be considered as a light test for a tolerably good 3 inch refractor. [I found it an easy object with this aperture in the Punjab sky] About 8′ *s. p* is the double star Struve 747, the components of which are rated by Lalande as 7, and 8, but I can now see the star steadily in full moonlight, with the naked eye.

σ. $5^h. 32^m$, *S* 2°. 40′ (1870). 4, 8, 7 : 84°.2, 60°.8 : 12″.5, 42″. There are 3 other stars *p* and an 11m star *p* 4. Smyth (Cel. Cycle) gives for the whole group :—mag's 4, 11, 8, 7, 8½ (D) 9, 8 and distances 12″, 12″.5, 41″.8, 211″.5, D—9, 8″½, D—8, 67″.8. Besides these there are several fainter *comites*. Struve counted 15 altogether with the Dorpat refractor of 9½ inches aperture. [I glimpsed some of these faint stars with 3 inches, 1874. With 3.9 inches I see 10 stars in all, four in each group, and two faint stars between the groups].

The small 11m *comes p* the primary is supposed to have brightened of late years, as it has been seen with 3 in, whereas Smyth (Cel. Cycle) says the observer need not " be very much annoyed with his instrument should he be unable to distinguish the minute *comes a* ; since it is so small a point of light that it escaped even the searching eye of ♅" . It is now plain enough with 3.9 inches. Ward has even seen it with 2⅞ inches.

The great nebula in Orion with the accompanying "trapezium" are so well known to telescopists as to need no notice here. An interesting description of this wonderful object will be found in Webb's Celestial Objects. I may add that I have never succeeded in seeing the 5th star with either of my instruments. I have however only examined the object on two or three occasions with the 3.9 inch.

PHOENIX.

A constellation, South of Sculptor. Its brightest stars are α of mag. $2\frac{1}{2}$, and β and γ of the 3rd magnitude. These do not rise above the English horizon.

Double stars.

3417. β. 1^h. 1^m, S 47°. 21′ ; $3\frac{1}{2}$, 11 : 17°.6 : 30″. The most southern of three stars γ , ν and β nearly in a straight line. I noticed in Nov. 1874 that ν (4m in Proctor's Atlas) was about equal to μ (5m). [*Comes* not seen with 3 inches, but the star is rather low in the Punjab sky.]

3463. 1^h. 39^m, *S* 44°. 49′ ; 9, 9 : 19°.6 : 20″ 'Points north of a star 9m'. Nearly following γ [Components equal, and fainter than the star to the North, which seems rather brighter than 9m, 3 inches, 1876].

251 Dunlop. θ. 23^h. 34^m, *S* 47°. 18′ ; 6,7 : 268°.8 : 4″.47. θ was rated 5m by Lacaille.

PISCIS AUSTRALIS.

A small constellation South of Aquarius, and Capricornus, and North of Grus. Its only bright star is Fomalhaut, a small 1st magnitude one. None of the others exceed the 4th magnitude. In 1874, I found the order of brilliancy of the stars in this constellation to be:— α, β, γ, δ, ε, ι, μ, η, θ, 1, 4, 15. In Proctor's Atlas γ and δ are marked 5m, and ε and ι, 4m. In 1752, Lacaille rated the principal stars as follows:— α, 1m. β, 4½m. γ, 6m. δ, 5½m. ε, 4m. ι, 6m. μ, 6m. and θ, 5½m. Heis gives β, 4m. γ, 5.4. δ, 5.4 . ε, 4m. ι, 5m. μ 5.4 and θ, 5.4.

A little south of 15 and μ, I noticed in 1874 two small stars of about 5½m, not shown by Proctor.

DOUBLE STARS. ETC.

α, Fomalhaut. $22^h. 51^m. 1^s$, *S* 30°. 15′.29″.

This fine white star is in a comparitively dark field. [With 3 inches, Nov. 1875 and Nov. 1876, I see no *near* comes, the nearest star being a *very* faint one about 5′ *preceding*. Another about 9½m to the South.] Smyth (Cel. Cycle) notes the 9½m star but makes no mention of the faint *comes* preceding. He calls the principal star "reddish," but I have never seen it of this colour. Fomalhaut is closely followed by a 6m star, Lacaille 9321 To this star I see a faint and distant *comes n.p*, 3 in. 1875.]

5296. A 2587. θ. $21^h. 41^m$, *S* 31°. 27′. Herschel gives 5½, 10 : 336°: 30″; also 5, 13: 334°: 25″ 'large star single with 320'. [*Comes* not seen with 3 in. 1875; must be much fainter than 10m. With 3.9 in, I found it obvious with averted vision; very small, about 13m. A faint and distant *comes* following. A little *preceding* is a small pair].

η. $21^h. 54^m$, *S* 29°.2; 5, 6: 120°±: 1″.8 (1874). Discovered by Burnham, who says "It is difficult to understand how so easy a pair has hitherto escaped detection. It is very plain with the 6 in, and I am quite sure an aperture of 3½ or 4 in. would show it well".

[Elongated ? with 133 on 3 in. Nov. 1876.]

5341. β. 22h. 25m, S 32°. 55′. 4½, 8 : 173°.3 : 29″.09 (1836). [White, reddish lilac with 3 in, and 3.9 in, 1875. A faint and distant *comes* about 11m to the South].

5356. 22h. 33m, *S* 28°.58′. 6, 7 : 159°.7 : 85″.31 (1836). 7 again double, 7,9 : 57°.6 : 4″.36 (1837). A little *s* of ε [Both yellowish. The close pair seen with 3 in, 1875].

5367, γ or 22. 22h. 46m, *S* 33°. 23′. 4½, 10 : 276°.8 ; 3″.59 (1835). 'The smallest point imaginable'. 'Bears strong illumination'. [Companion not seen with 3in, 1875. Steadily seen with 166 on 3.9 inches, Nov. 1876. Glimpsed with 104. The *comes* seems about equal in magnitude to a 10m star *s.p.*].

SAGITTARIUS.

One of the constellations of the Zodiac, situated South of Aquila, and East of Ophiucus and Scorpio.

Its brightest stars are ϵ, σ (2m) δ and ζ (3m) which form a trapezium. σ is said to be slightly variable. α is at the present time only of the 4th magnitude, so that it may possibly be variable also. The Milky Way in this constellation is especially brilliant.

Double stars.

5007. 17^h. 54^m, *S* 23°.2′. Triple; 7, 11, 8: 21°.1, 211°.3 : 5″, 15″. Involved in the great 'trifid nebula' 20 Messier. A litttle *n.f* the 5m star, 4 Sagittarii.!

5035. μ'. 18^h. 6^m, *S* 21°.5′. Quadruple. 3.5, 9.5, 10, 13: 315°, 114°.5, 267°$\pm$: 40″, 45″, 18″. [The two larger *comites* readily seen with 3 in, but not the fourth star]. Smyth rates the 4th star 16m, and only 10″ from the primary.

5073. A 2179. ν'. 18^h. 44′, *S* 22°.57′ (1830); 6, 10: 60°.3 : no distance given. North of

5107. A 2240. Dunlop. 226. β'. 19^h. 13^m, *S* 44°. 42′ (1870). 4½, 8: 78°.6 : 29″.09, (1835). [8 lilac, 3 inches, 1876. A distant small star *follows*, a little *South*].

5112. 19^h. 16^m, *S* 18°. 23′ (1870). Quadruple; 8, 8, 8, 12 : 229°.6, 50°±, 180°± : 18″, 20″, 25″. [The position angles, given by Herschel as above, would place the three 8m stars nearly in a straight line, but I find an observation on this object as follows:—9th July 1874, 3in. refractor, 5112 *h* Sagittarii. The three 8m stars form a neat triangle, and an interesting specimen of a coarse triple star, 12m not seen, (50° is perhaps a misprint for 150°). The three stars are not exactly equal (as remarked by "F. R. A. S" in *English Mechanic*), the most Northern being the brightest, next the *preceding* star, and the *f* star the faintest. The object follows, and nearly forms an equilateral triangle with ρ^1 anb ρ^2 Sagittarii]. Burnham (in 1873) failed to see the 12m star, and says that the angle 50°

should be 140°, a mistake of 90° having been made by Herschel*

54 Sagittarii. $19^h. 33^m$, *S* 16°. 35′. 5½, 8: 42°.8: 28″. Webb says "yellow and pale lilac. Fine field of minute stars". [The *p* of two stars of nearly equal brilliancy. South, a little *p*, is a fine pair about 8m and about 8″ distant, seen with a power of 40 on 3 in. This, according to Burnham, was measured by Sir J. South in 1827 : P 237°.3 ; D , 10″.67 ; mags. 8, 8.2.*].

3 Sagittarii . $17^h. 40^m$, *S* 27°. 47′ was discovered to be variable in 1866 by M. Schmidt of the Athens Observatory, who makes it period about 7 days, and its variation 4m to 6m. The latest period given by Schmidt is $7^d. 0^h. 17^m. 42^s.5$. [On the 15th July 1876 I found it about equal to 4 Sagittarii (5m Heis), but slightly less than 45 (*d*) Ophiucii (5m Heis)].

Sagittarii. $17^h. 57^m$, *S* 29°. 35′ is also said to be variable by Schmidt from 5 to 6.5m in about

* English Mechanic. 3rd December 1875.

7.6 days. [On 15th July 1876, I found it about equal to a 6m star *n.f* (6.5 Heis)].

Red Star. $19^h. 27^m$, *S* 16°. 39′. A little preceding the double star 54 Sagittarii. It was rated 7m by Sir J. Herschel who describes it as "remarkably red, high scarlet, or good ruby colour". It is 8m in Harding's Atlas. [About 7m, 1875 and 'fiery red', even redder than Antares. The most Southern of 3 stars forming a curved line].

A little *n. f* π Sagittarii (3m) are 3 small variables R. S and T Sagittarii discovered by Pogson. Of these S is considered by Schiaparelli to be identical with a Temporary Star which appeared in the year 1690. R and S are near a 5m star (Proctor) 43 (*d*, Lalande) which may also possibly be variable, as it was rated 4m by Harding, and 6m, 8m by Lalande. Argelander and Heis have rated it 5m. Its place for 1880 is R.A. $19^h. 10^m. 37^s$, *S* 19°. 25′. 18″. [In June 1875, it seemed to me about 6m, being then much fainter than ρ^1 (5m) but brighter than ρ^2, and a 7m star *following*.

In Nov. 1875 it seemed rather brighter, or about 5½m. A little less than υ, Oct. 1876].

σ 18h 48m, *S* 26°.27′. 3, 9½: 243°.5: 309″; 'ruddy, ash coloured'—Smyth (Cel. Cycle), who says "a small star on a similar rhumb to the *s. f.* makes the whole form a triangle with the apex to the North", [so I see it with 3 inches,] and adds; "This star has been placed among the variable ones under a probability of its varying from the 2nd to the 4th magnitudes; but its low altitude might occasion apparent changes *. Ptolemy, Ulugh Beigh, Bradley, De Zach and Mayer have classed it 3; Flamsteed 3½; Tycho Brahe′ and Hevelius 4; but Bode makes it 2m, and Lacaille and Pigott 2½". Heis rates it 2.3. [In August 1876, I found it about equal to ε Sagittaii (3.2 Heis), but less than λ Scorpii. A little brighter than ζ Sagittarii. Heis rates λ Scorpii incorrectly at 3m. It is now about 2m. Heis' estimate however of 2.3 for σ Sagittarii is about right].

*I have seen the bright (1m.) star Achernar (α Eridani) close to the Southern horizon in the Punjab as a star of about 4m. On a finer night however, I have seen it 2½ or 3m. Its Meridian altitude at my station was about 2°.

NEBULÆ AND CLUSTERS.

3718. IV. 41. 20 Messier. 17^h. 54^m, *S* 23°. 2′ (1870) *North following* 4 Sagittarii. The 'trifid nebula' of Sir J. Herschel. It surrounds the triple star 5007 *h* Sagittarii (*ante*).

3722. M 8. 17^h. 56 ^{m}, *S* 24°. 21′ (1870) *South following* 4 Sagittarii (4m). The star 9 Sagittarii (7m) is involved in the nebula 'followed by a great cluster VI. 13 which with the nebula fills many fields'. Webb says 'visible to the naked eye'. [Plain to the naked eye in the Punjab. A glorious object even with 3 in: the cluster *f* being beautifully seen, but the nebula is milky looking, though several small stars are visible in it. The star 7 Sagittarii *preceding* is avery wide double, the companion being also double. The cluster lies midway between two 8m stars].

3753. M. 22. 18^h. 29^m, *S* 24°.1 (1880). "Globular; very bright, large and compressed; 7′ diameter. The stars are of two magnitudes, viz:—15-16 and 12m, and what is very remarkable, the largest of the latter are visibly reddish; one in particular the largest of all *s f* the middle is decidedly a ruddy star, and so I think are all the other large ones". Herschel also says, 'comparing its place with *h* 2015 a suspicion of proper motion arises'. About $2^\circ\frac{1}{2}$ *north following* λ Sagittarii (3m). Webb says the largest of the component stars are 10 and 11m. [The larger stars well seen with 133 on 3in. They seem much fainter than 10 and 11m. Herschel's estimate of 12m is more correct. The greater portion of the cluster is nebulous with 3 inches. It is 7786 in Lacaille's Catalogue, and comparing its place as found by Lacaille in 1752 with that of Herschel for 1830, the suspicion of proper motion does not seem to be confirmed, as the positions (brought up to the same date) differ by only 2^s in R.A. and 27″ in N. P. D.].

3798. M. 55. $19^h. 31^m$, *S* 31°. 16′. "Globular; a fine large, round cluster; 6′ diameter; all clearly resolved into stars 11, 12, 13m; does not come up to a nipple". About 9° following ζ (3m) and *North preceding* a 5m star. [Glimpses of stars seen in it with 40 on 3in, 1875; will not bear a high power with this aperture; very much like, but not so bright as 13M in Herculis. The larger stars seem fainter than 11m.].

SCORPIO.

This well known constellation contains many interesting telescopic objects. A portion of it only is seen above the Southern horizon in England during the short summer nights, but the whole of the curved tail is visible in the Punjab. Its leading brilliant Antares is a fine red star of the 1st magnitude.

Double stars, etc.

(Antares.) 16^h. 22^m, *S* 26°. 10′; 1, 7½ : 275°.7 : 3″.7 (1864). Binary? Large star 'fiery red', small star green, a beautiful combination. The observations would seem to prove that it is a binary pair; Dawes in 1864 found the position and distance as given above, whereas Wilson in 1873 found the position angle 268°.6, showing an angular motion of 7°.1 in 9 years, indicating a probable period of about 450 years. The fact of its being double does not seem to have been known to Sir John Herschel when at the Cape of Good Hope.

4839 . 12 Scorpii. $16^h. 5^m$, S 28°. 6′ ; 6½, 10 : 81°.7: 4″.62 (1834).

4843 . σ . $16^h. 14^m$, S 25°. 19′; 4½,9 ; 274°.3 : 15″. Webb gives 4, 9½ ; 271°.6 : 20″. [*Comes* small with 133 on 3in., 1874.] Lacaille rated σ at 3½m.

4886. Dunlop 209 . $16^h. 40^m$, S. 36°. 39′ ; 7½, 8½ : 147°.3 : 23″.21 (1837). Herschel says 'surely this star must have changed in distance' [*n. p* μ' Scorpii.

4962. B. 6125. $17^h. 26^m$, S 32°. 30′; 6½, 13 : 102°.3 : 8″. Multiple star, about 5° North of λ Scorpii. Herschel says "15 or 20 more stars 13m clustering round the large star. Very curious. The great star occupies the centre. A very remarkable object".

A diagram should be made of this group with a large instrument, to ascertain if their relative positions alter. If the stars are relatively close—as seems probable—some motion must result. [Several of the small stars (5 or 6) seen with 133 or 3 inch, 1875].

Red Star. $16^h. 32^m$, S 32°. 7′. 8m. A little *n. p* ε (3m). Herschel describes it as "deep red, like a drop of blood."

Red Star. $17^h. 31^m$, S 41° 33′. 8m. About midway between θ and ι Scorpii. Sir J. Herschel describes it as a "beautiful ruby red".

Clusters and Nebulæ.

3652. Dunlop 499. $16^h. 44^m$, S 41° 34′. "A fine bright large cluster; 10′; stars 10 to 13m. Place of a double star 5m, the *p* but one of 7 bright stars in the middle". [Just North of ζ^1 and ζ^2 Scorpii. Very bright. Visible to the naked eye in the Punjab as a hazy star of 4 or 4½m. The component stars well seen with 3 inches. A beautiful object for a small telescope].

3654. Dunlop 520. $16^h. 46^m$, S 39°. 17′, (1870). "A fine large rich cluster; 12′; irregular figure; stars 9 to 12m; place of a red star 9m in centre". *s.f* μ^2 Scorpii. It is No 7038 in Lacaille's Catalogue. [Elongated N and S. With 3in. the

9m red star seems *just outside* the cluster *s.f* (July 1875)].

3699. 6 Messier. $17^h. 32^m$, *S* 32° 9′. "Chief star 7m of a fine large discrete cluster of stars, 10, 11m; one star is 7m; one is 7.8. Fills field". [*Following* a little *North* of the multiple star 4962 *h.* From the position given by Herschel, this cluster should be in the same low power field with the following—612 Dunlop. With 3in. however I see only *one* cluster, the appearance of which agrees better with Herschel's description of 612 Dunlop than that of 6 M.]

3702. Dunlop 612. $17^h. 35^m$, *S* 32°. 15′. 'Coarse, very large cluster. Stars 8 to 12m. Not rich. Stars in zig-zag lines'. [The most Northern of the brighter stars reddish, 3 in., 1875. The cluster 6 M above should be in the same low power field with this cluster, but with 3in, I can only find *one* cluster near the place, the *appearance* of which agrees perfectly with Herschel's description of Dunlop 612, the stars being from 8 to 12m, and 'in zig-zag lines' The *position* of the cluster however agrees more closely with that given for

6 M than for Dunlop 612. Are the clusters identical?].

3706. Dunlop 597. $17^h. 41^m$, *S* 34° 49′ (1870). 'A very fine large rich scattered cluster of stars 12 to 13m.' [Precedes the cluster 7M. This seems to be identical with 3710 (below), as with 3in and a large field I see only *one* cluster].

3710 . $17^h. 45^m$, *S* 34° 47′ (1870). 'A brilliant close cluster of about 60 stars, 7 to 12m, which fills the field.' [Visible to the naked eye. Beautifully seen with 3 inches. Possibly identical with 597 Dunlop (3706) above, as with 3 inches, I can see no trace of a second cluster near the place].

SCULPTOR.

A constellation, South of Cetus and Pisces, and North of Phœnix. It includes no star brighter than the 5th magnitude, but contains a few objects of telescopic interest. The Southern pole of the Milky Way is situated in this constellation, a little North of the star

Double Stars, etc.

3436. $1^h. 21^m$, *S* 30°.51′; 7, 10 (or 7½, 9): 123°.7 : 13″. 'Fine star, pale yellow, pale blue'. In a space very barren of bright stars *f* α.

3442. $1^h. 26^m$, *S* 26°.3′; 6½, 10 : 208.3 : 30″. About 3° *s.p* ε (5m). Not in Harding's Atlas.

3461. A. 191. ε. $1^h. 40^m$, *S* 25°. 42′; 6, 10: 69°.6 : 5″.53. "Delicate". 5m in Proctor's Atlas, about 9° South of τ Ceti (3m). [Companion glimpsed with 3 inches, 1875; difficult. A faint *comes* to the South].

Red star. $1^h. 21^m$, *S* 33°. 13′. 6m. "A most beautiful orange". It will be found near a number of 6m stars *s. f.* α. Dr. Gould of Cordoba says this is one of the most intensely red stars he knows in the sky, and calls it "brilliant scarlet".

SERPENS.

A long narrow constellation. The portion

South of the Equator is situated between Aquila and Sagittarius on the East and Ophiucus on the West, and contains but few objects of interest for a small telescope, the most important being;

ν. $17^h. 14^m$, *S* 12°. 43′; 4½, 9: 31°.3: 51″. The colours according to Webb were, in 1851, 'Silvery, and native copper' and he considered the companion small for 9m. [*Comes* well seen with 40 on 3 inches in strong twilight, Nov. 1876. A distant faint star *n.p* mentioned by Smyth (*Cel. Cycle*) also seen].

NOTES ON SOME SOUTHERN STARS POSSIBLY VARIABLE.

Ces variations des étoiles sont bien dignes de l'attention des observateurs curieux....Un jour viendra, peut-être, où les sciences auront assez d'amateurs pour qu'on suffire à ces détails.— LALANDE, *quoted by* SMYTH.

The following are some stars South of the equator which may be suspected to be variable. Many of them are underlined in Harding's *Atlas, Novus Cœlestis* (1822), which seems to indicate that, although observed by Harding, they were not found by him in the Catalogues to which he referred. Most of these stars are missing in Lalande's large Catalogue, which includes the great majority of the stars shown by Harding. The remaining stars in the list have been for other reasons suspected of variation. The objects are given in order of Right Ascension, and their places noted with reference to the nearest naked eye star.

The positions are for 1880.

(1)

Lalande 1028—29—30.

R A. 0^h. 34^m, *S* 8°. 40

Rated once as 6½m, and twice 9m by Lalande. It is 7m in Harding's Atlas (no line), and lies about 3°. *n.p* ϕ' Ceti (6.5 Heis). [I found the star about 7 or 7½m. Nov. 1876.]

(2)

About R A. 3^h. 9^m, *S* 6°. 20′.

Not in Lalande's Catalogue, 7m Harding. About 3°, nearly due N of ζ Eridani (3m). [I found it about 6m. Feb. 1876, and brighter than a star *n.f*]. Heis rated it 6m, and identifies it with III. 147 of Weiss' second Catalogue. [Nov. 1876, not very much brighter than the star *n.f*]. Both stars are underlined in Harding's Atlas.

(3)

About R A. 3^h. 10^m, *S* 6°.10′.

Closely *n.f* (2). 7m Harding. Not in Lalande's Catalogue. It is a red star, and is No 26 of Chamber's Catalogue of red stars (*Descriptive Astronomy*) where it is given as 7m.

[I found it about 6½m. in February 1876, and less than the star South of it].

(4)

About R.A. $4^h. 28^m$, *S* 8°. 29′.

Closely *p* a little *n* of 47 Eridani (6.5 Heis) 7m Harding. Not in Lalande, in which 47 Eridani is also missing, and also an 8m star *s.f* 47, shown by Harding. About 44′ south of 47 Eridani, Heis gives a 6m star, which is not in Harding's Atlas, nor in Lalande's Catalogue. It is IV 585 of Weiss' second Catalogue.

[In February 1876 and Jan. 1877 I found Harding's star about 6½m. Heis' 6m star fully equal to 47 Eridani].

(5)

About $4^h. 32^m$, *S* 30°.38′.

Closely *n.f* υ^7 (Proctor and Lacaille) Eridani (υ^3 Harding; υ^2, Heis) (52 Flamsteed). 6m in Harding's Atlas. Not in Lacaille's Catalogue.

[There is no star of 6th magnitude now (Feb. 1876) in this position, but with 3 inch refractor I see, close to the place, a small star about 10½ or or 11m, which may be variable and should be watched. In the field *s. f* v^7 is a 6m star, and about the same distance (or a little further off) *n. f*, a star about 6½m. [Both these latter stars were observed by Lacaille, and are shown by Harding.]

(6)

About $4^h. 39^m$, *S* 18°. 36′.

6m Harding. Not in Lalande.

It is situated 20′ due North of the 6m star Lalande 8951, which lies 1° 20′ *n.f* 54 Eridani (5m Heis).

[In Feb. 1876, I found it about 9m (!), and a little fainter than another star a few minutes to the North].

(7)

About $4^h. 44^m$. *S* 15°. 58′.

7m Harding. Not in Lalande.

27′ North of 60 Eridani (1498 B.A.C.) 6m. [With 3in. I found it about 7 or 7½m. Feb. 1876. It is the *f* star of four forming a trapezium. Within this trapezium are 3 faint stars, 3in. 1876].

(8)

About $4^h. 58^m$, *S* 21°. 14

6m Harding. Not in Lalande's or Heis' Catalogues.

It is situated a little *n.p* ε Leporis, and between that star and 9506 Lalande. (5m) (Heis 1553 B. A. C. 5.6m). Closely south of this 6m star, Harding marks a 9m star, and *n. p.* three more 9m stars, forming a curved line. None of these small stars are given by Lalande, but probably their magnitudes are overrated by Harding.

[In Feb. 1876 with 3in, I found Harding's 6m star a little brighter than the 9m star south of it, but *less* than several 8 m stars (Harding) following. It has a small companion *f* at about 1′± distant.

Harding's 9m stars are more like 10m].

(9)

9552 Lalande. 1 (Fl) Leporis.

$4^h. 58^m$, *S* 22°. 58′.

s.p. ε Leporis (4.3m).

A suspected variable. Rated 6½m by Lalande, and 6m in Harding's Atlas, but not in Argelander's *Uranometria* nor in Heis' Atlas*.

[In Jan. 1876 I found it about 6½m, and *just* visible to the naked eye of a clear night. It is of

* *Nature* 27 May 1875.

an orange tint in the telescope, and is *preceded* at about 3′ distance by a small star 10½ or 11m].

(10)

λ Eridani.

5ʰ. 3ᵐ, *S* 8°. 54′.

Closely *s.p* Rigel (β Orionis).

This is a suspected variable, and has been classed by different observers from 4m to 6m. It was rate 4, 4½ by Lalande, 5m by Harding, and was estimated at 4.97 by Sir J. Herschel at the Cape of Good Hope. Heis rates it 4m.

[In March 1875, I found it about equal to ω Eridani: brighter than ψ Eridani, but not quite equal to τ Orionis (4.25 Herschel, 4m Heis). Its proximity to so bright a star as Rigel may perhaps have something to do with the uncertainty as to its exact magnitude].

(11)

About 1° *n.f* θ Leporis (5m) is a star of about 6½m, which is not in Lalande's Catalogue, nor in Harding's Atlas. It is VI. 58 of Weiss' first Catalogue, observed by Bessel in 1825 as 6m.

It is 6.7m in Heis Catalogue. No observations of it however were recorded by d'Agelet, Lamont and other observers, and it is possibly an irregular variable of long period, (*Nature*, May 20th 1875).

[The star is of a reddish hue, and is in the same low power field with, North of, and a little *p* the 7m star Lalande 11,778. The position of this latter star is for 1875. RA. $6^h. 4^m. 38^s$, *S* 15°. 2′. 1″].

(12)

Lalande 12104.

$6^h. 14^m$, *S* 2°. 53′.

A fine orange star, the colour of which was discovered by Birmingham, who suspected the star to be variable. It was rated 6m by Lalande, Harding and Heis. It has an 11m *comes* nearly *f*.

[I found this star just visible to the naked eye in the Punjab. Feb. 1875].

(13)

22 Canis Majoris.

$6^h. 57^m$, *S* 27°. 45′.

This star which is identical with σ Canis Majoris is situated between δ and ϵ. It is red and is possibly a variable. It is 4m in Lacaille's Catalogue and

Harding's Atlas, and was also estimated 4m. by Heis, but it was once rated 5m. by Argelander, and this is the magnitude, assigned to it in the Washington General Catalogue.

[In Feb. 1875 I estimated it as 4½m.

In Jan. 1876, I found it very slightly brighter than ο′ (5m Heis). It may be variable like many other ruddy stars, and should be watched by Southern observers].

(14)

About $7^h. 5^m$, S 15°. 29′.

7m Harding. Not in Lalande's Catalogue.

A little *f* γ Canis Majoris (4½m), and 30′. *s.f.* the 7m star Lalande 13925.

[I found it about 7½m. Feb. 1876. and equal to LL 13925].

(15)

27 Canis Majoris.

$7^h. 9^m. 22^s$, S 26°. 8′. 58″ (1880).

Near δ. This star was rated 5½m by Lacaille, 4m by Harding, 5m Argelander, 6m Behrman, 6.5 Heis, and is 7m in Flamsteed's Catalogue.

[In 1874, 1875, and 1876, I found it about 5½m,

and much inferior to 28 which Harding shows as 5m. It may possibly be variable from 4½ to 6½m. with a long and irregular period.]

(16)

$7^h. 23^m$, *S* 1°. 39′.

7m Harding. Not in Lalande.

About 28′ *n p* the 7m star LL 14,645. Observed as 4½m by Rumker in 1822, and suspected by Olbers in 1824 to be a remarkable variable.

It was rated only 8.9m by Fellocker in the Berlin Academy charts, and is not given in Argelander's *Uranometria.** In Heis' Catalogue it is rated 6.7m, and identified with VII. 669 of Weiss' second Catalogue.

[I found it, March 1875, about 7m; fainter than 25 Monocerotis (6m). In Jan. 1876, and Jan. 1877, it seemed about 6½m, being less than 25, but brighter than two 7m stars *following*].

(17)

$7^h. 23^m$, *S* 9°. 51′.

7m Harding. Not in Lalande's Catalogue.

It lies about 20′ *n. p* the 6m star LL 14,599

* *Nature*, Feb. 4th. 1875.

(6m Harding) (and about 3° preceding, a little South of the 4½m star 26 Monocerotis.)

This latter star was suspected to be variable by Birmingham, in April 1875, who found it to be of a "red orange" colour, and mag. 7.3. In June of the same year Tebbutt at Windsor. N. S. Wales estimated it 6.3m. About 40′ *preceding*, Harding underlines an 8m star which is not in Lalande's Catalogue.

[In Feb. 1876, I found Harding's 7m star about 7m, and half a magnitude fainter than LL 14,599 which seemed about 6½m. Harding's 8m star also visible with opera glass. Jan. 1877; about the same relative brilliancy. About 40′ *n.f* Harding's 7m star is the star Lalande 14,658, rated 8m by Lalande and Harding. This star is now (Jan. 1877) about 6½m, and slightly brighter than LL 14,599].

(18)

About $7^h. 28^m$, *S* 19°. 3′.

It is situated 48′ *n.f* the 6m star Lalande 14,893 (Puppis). 7m in Harding's Atlas, but not in Lalande's Catalogue.

[About 7m Jan. 1877; less than LL 14,893, but brighter than LL 14,808, to the South of it.

(19)

Lalande 19,662 (Sextans).

$9^h. 58^m$, S $8^\circ. 59'$.

About $2^\circ\frac{1}{2}$ *n* of and a little *p* λ Hydrae (4m).

[Rated 4½m by Lalande, and 5m in Harding's Atlas, but observed by me in April 1875 to be only 7m.]

It is 6.7m in Heis' Catalogue, and was rated 7.8m by Lamont in 1845, but does not seem to have been observed by either Argelander, d'Agelet, Bessel or Santini (*Nature*, May 13th 1875).

It is closely *preceded* by a 7½m star, LL 19,646.

(20)

$10^h. 11^m$, S $18^\circ. 49'$.

7m Harding. Not in Lalande's Catalogue.

It lies about $3^\circ\frac{1}{2}$ *s. p* μ Hydrae (4m), and between the 6m stars, LL 19,880, and LL 20,137.

About 10′ *f* a little South is the 9m star LL 19,998.

[I found Harding's star about 7m. March 1876.]

(21)

10 Hydrae.

$11^h. 2^m. 45^s$, S $27^\circ. 25'$. (1876).

s. f χ^1 and χ^2 Hydrae.

A suspected variable by Gilliss, (*Nature*, 20th April

1876) [On 15th May 1876, I found it about 5m, and slightly less than χ^1. Much brighter than a star *s. p* (7m Lacaille and Harding)].

[Closely *s.f* Harding shows a 6m star. There is a star near this place at the present time, but not so bright as 6m.]

(22)

11ʰ. 12ᵐ, *S* 3°. 15′.

7m Harding. Not in Lalande.

About 24′ *s*, a little *p* ϕ Leonis (4½m).

About 7′ *n.f* is an 8m star LL 21,530.

[In March 1876, I found Harding's 7m star about 6m; nearly equal to 69 Leonis, and brighter than 66 (between 61 and 69), which Harding rates 6m, and Heis 6.7].

(23)

31 Crateris. LL 22,591-92.

11ʰ. 55ᵐ, *S* 18°. 59′.

Between η Crateris and ϵ Corvi.

This star is not shown in Proctor's Atlas. It is 5m in Harding's Atlas, and was rated by Lalande once as 4½m, and once as 6m. Heis rates it 6m.

[In April 1875, I found it about equal to Crateris (Heis 6m.) Not equal to ζ, which is about 5m. In April 1876. it seemed very nearly equal to ζ. The difference between ζ and η however does not appear to me to be a full magnitude as given by Heis].

(24)

Lalande 23,228–9.

$12^h. 19^m$, *S* 10°. 56′.5.

Rated 7m by, Lalande in 1795, and 5½m in 1798. It is 7m in Harding's Atlas, and was estimated 6.7m by Lamont, and only 8m by Steinheil. It is not given by Bessel or Santini (*Nature*, May 6th 1875) Heis rated it 6m.

It is situated about 5° *s. p* the 5m star χ Virginis and about equidistant from that star and δ Corvi. About 30′ *s. p* is the star Lalande 23,186, 6m Lalande and Harding, but not in Heis' Catalogue.

[I found Harding's, 7m star about 6m in March 1876, and much brighter than LL. 23,186. Not quite equal to a 6m star (Harding about 2° *s. p*)].

(25)

Lalande 23,726 (Corvus).

$12^h. 33^m$, *S* 13°. 12′.

Rated 7½m by Lalande, 8m by Harding and Bessel, but seen by Heis as 5m.

It lies about 4° *n.f* δ Corvi.

[In June 1875, I found it about 8m. March 1876 it seemed about 7 or 7½m].

About 40′ North, Harding marks with a line a 7m star which is not in Lalande's Catalogue. [7 or 7½m March 1876].

About 1° *n.p* LL 23,726, is the double star, Struve 1669 (Lalande 23,675-76. 7½, 7m). 7m in Harding's Atlas, but rated 5m by Heis. [It seemed to me about 5½m in June 1875.]

(26)

About $12^h. 49^m$, *S* 4°. 14′.

About 50′ *s.f* the 7m star Lalande 24,016 ($12^h. 47^m$, *S* 3°. 34′) and 3°½ *p* a little *n* of Virginis (4.5m).

It is marked with a line (7m) by Harding, but is not given by Lalande. (About 25′ North of it Harding shows another 7m star (no line) which is also missing in Lalande's Catalogue).

[I found it about 7m and brighter than the star *n* of it, March 1876].

(27)

13^h. 0^m, *S* 13°. 34′.

About 2°½ *n.p.* 53 Virginis.

Marked 7m with a line by Harding. Not in Lalande.

[I estimated it about 7½m in April 1876.

In May 1876 I found it considerably fainter than the following (28).]

(28)

13^h, 3^m, *S* 15°. 46′.

About 40′ *s. p* 53 Virginis (5m).

7m Harding. Not given by Lalande or Argelander. It is rated 6.7 by Heis, and identified with 4396. B. A. C.

[About 6½m, April 1876].

About the same distance *s.f.* 53 Virginis is a star suspected by Olbers in 1797 to be a remarkable variable. It is not in Harding's Atlas.

[In May 1876, I found it about 9m, and equal to LL 24,421. (Olber's star *c*), but brighter than LL

24,597. (Olber's star *d*)].* In July and August 1876, Tebbutt at Windsor N. S. W. estimated it at 8½m.

(29)

ε Libræ.

$15^h. 18^m$, *S* 9°. 53′.

This star is 4m. in Lalande's Catalogue, and also in Harding's Atlas. [In May 1875, I found it about 5½m., being then rather fainter than 37 Libræ (5½ Lalande, 6m Harding)]. Both stars were rated 5m by Heis.

(30)

About $17^h. 10^m$, *S* 12°. 58′.

7m Harding. Not in Lalande's Catalogue.

1° *p* a little South of ν Serpentis (4½m).

[I found it about 8m. Aug. 24th 1876].

(31)

$18^h. 1^m$, *S* 17°. 8′.

6m Harding (underlined). Not in Lalande's Catalogue, nor in Heis' Catalogue. It lies about 4° *n. p* μ Sagittarii.

* *Nature* April 13, 1876.

(32)

North preceding the 6½m star 29 Sagittarii, and *n.f* μ Sagittarii, is a scattered group of 7m, and 8m stars (Harding). Among these are seven 7m stars which Harding marks with a line, and all are missing in Lalande's Catalogue. The variables U and V Sagittarii, and the cluster 25 Messier are included in the *p* portion of this group, which is situated in a very luminous part of the Milky Way.

(33)

$18^h. 50^m$, *S* 22°. 4′ (1870).

This is No. 233 of Chamber's Catalogue of Red stars (*Descriptive Astronomy**). It is not in Lalande's Catalogue, nor in Harding's Atlas. It lies between 33 Sagittarii, and ν^1 and ν^2 Sagittarii.

33 Sagittarii is also missing in Lalande, but is 6.7m in Heis' Catalogue. [I found it about 6½m Oct. 1876 and equal to 30 Sagittarii. Slightly brighter than 31 Sagittarii].

(34)

ι Aquilæ (41 Fl).

$19^h. 31^m$, *S* 1°. 33′.

Lalande 3½, 4 (37142-43). d′ Agelet 6, 4½. 5m

* 2nd Edition.

Harding. About 5° *s. p.* the variable η Aquilæ, South of Altair. Heis rates it 4.5.

[I found it about 4½, or perhaps nearer 5m at the end of August 1875. Brighter than ν, Oct. 1876, but slightly less than 12 Aquilæ].

(35)

A pair of 7m stars in Sagittarius. About $19^h. 41^m$, *S* 17°. 23′. A little *following* a line joining 55 Sagittarii (5m Heis) with 57 Sagittarii (6m Heis) marked, with a line, by Harding, but not in Lalande's Catalogue. [About 7m. Aug. 1876].

The star 57 Sagittarii is shown 5m by Harding and 56, 6m. These magnitudes are transposed by Heis. 57 is missing in Lalande's Catalogue.

[In Aug. 1876, I found 56 Sagittarii about 1mag. brighter than 57. A little *s.f.* 57 is a star about ½ a mag. fainter than 57. This is Lalande 37,873, 8m, also shown as 8m in Hardings Atlas, but it seems to be now about 6½ or 7m].

(36)

ξ^1 Capricorni.

$20^h. 5^m$, *S* 12°. 45′.

Rated 6, 6½ and 7½ by Lalande. 6m in Harding's

Atlas. It lies about 15′ *n. p* ξ^2 Capricorni (6m Harding's and Heis, 6m and 7m Lalande).

About 17′ *n* of ξ^2 Harding marks with a line an 8m star, which is not in Lalande's Catalogue.

[In Oct. 1875, I found ξ^1 about one magnitude less than ξ^2. About the same difference Aug. 1876.]

(37)

V Capricorni.

Lalande 38.839.

$20^h. 10^m$, *S* 21°. 41′.

About 2°½ *s.p* σ Capricorni, and *n.p* 4 Capricorni (6m). This star was rated 6½m by Sir J. Herschel, who says "A fine ruby star. This is perhaps the finest of my ruby stars". 7½m Lalande. 7m Harding.

[I failed to see this star with an opera glass in July 1875, and with 3 inch refractor, found it only 8½ or 9m, and fiery red. If Herschel's estimate of 6½m was at all correct, the star must certainly be variable to the extent of 2 magnitudes. I propose to call this star V Capricorni. Closely following a little *North* is a fainter star about 10 or 10½m.]

[About 1° *n* a little *f* 4 Capricorni, I observed in

July 1875, a 7m star, which is not in Lalande's Catalogue or Harding's Atlas.]

[In Nov. 1876. I found the red star about 8½m and much fainter than the 7m star above alluded to. Colour fiery red, 3in refractor. Visible with Dicratic Binocular by Browning].

(38)

$20^h. 37^m$, *S* 12°. 4′.

About 2°½ *s.p* ε Aquarii (4.3m.)

6m Harding. Not given by Lalande or Heis.

[I found it about 7m Aug. 1876].

(39)

$20^h. 42^m$, *S* 5°. 10′.

About 20′ *n.f* 3 Aquarii (4½m.)

7m Harding. Not in Lalande's Catalogue.

[About 7m August 1876].

(40)

$20^h. 59^m$, *S* 17°. 43′.

Star 6m Harding. Not in Lalande's Catalogue. About 30′ *s.f* the 4m star θ Capricornii.

With reference to this star, Heis, who rates it

6.7m, remarks "Stella ab Hardingio observata a me 17. Aug. 1852 et 19. Aug. 1865 nudis oculis videbatur sed 17. Aug. 1863. et 23 Sept. et 27. Sept. 1870 non conspicua erat. Stella fortasse variabilis est."

[On 10th Aug. 1876, I found this star, (with opera glass) about 6m and equal to 20 Capricorni. A little brighter than 21 Capricorni, which Heis rates as 6.7m and identifies with 7282. B. A. C.

About the same magnitude, 18th Oct. 1876].

(41)

33 Capricorni.

R A. $21^h. 17^m$, *S* 21°. 21′.

Red and a suspected variable.

About 1°. 40′ *n.p* ζ Capricorni (4m).

It was rated 6m by Lalande (41,543), 5m by Harding 5.6 by Argelander, and 6.5 Heis. Chacornac remarked that it was "sometimes brighter and sometimes fainter than a star of the seventh magnitude near it".

[I estimated it 6½m, August 1875.

In Aug. 1876, it seemed about 6m, and slightly brighter than 35 Capricorni. About the same magnitude Oct. 1876].

(42)

21^h. 39m, *S* 13°. 23′.

About midway between 45 (6m) and λ Capricorni (5½), 6m Harding, but not in Lalande's or Heis' Catalogue. [I found it, 24th Aug. 1876, about 7m and less than 44 Capricorni. About equal to 50 Capricorni, South of λ.]

(43)

Lalande 42,672-73-74.

21^h. 47^m, *S* 4°. 53′.

This star which is not in Proctor's Atlas is shown 5m by Harding, and was rated 5,6 and 6½ by Lalande. Heis rates it 6m, and identifies it with 7628 B. A. C.

[I found it about 6½m in Oct. 1874].

About 1° South of this star Harding marks with a line a 7m star which is not given by Lalande.

(44)

30 Aquarii.

21^h. 56^m, *S* 7°. 9′.

[This star which is marked 5m by Harding and

Proctor and rated 4½m by Lalande, I found about 6m in Oct. 1874, being then about equal to 51 (5½ Lalande) but decidedly less than κ Aquarii (5, 5½, 6, 6½ Lalande). Either 30 or κ Aquarii is probably variable].

Heis rated both stars as 5.6m.

(45)

Lalande 43,239-40.

$22^h. 4^m$, *S* 4°. 54′.

This star is not in Proctor's Atlas, but is shown 5m by Harding, and was rated 5m, and also 7m by Lalande. It is not in Heis' Catalogue.

[In Oct. 1874, I found it about 6½m].

About 25′ north, a little *p.* is the star LL. 43,226-27, rated 7, and 5½ by Lalande. It is 6m in Harding's Atlas, and also in Heis' Catalogue (7720. B. A. C.). [I found it, Aug. 1876, equal to the star South of it].

(46)

Lalande 45,145.

$22^h. 59^m$, *S* 17°. 33′.

7½m Lalande. 7m Harding, (underlined).

Closely *North* of and a little *f* a 6m star

Harding) which lies about 3° *s. f.* δ Aquarii. This latter star is Lalande 45,137 (7m) rated by Heis 6.7. The pair are situated in a vacant space in Harding's Charts.

[With 3in. refractor in Nov. 1876, I found it about 1 mag. less than the star South of it. About 90″ to the North of it, a little *p*, is a small star about 11m].

(47)

$23^h. 9^m$, *S* 10°. 18′.

A little *s.p* ψ^1 Aquarii. 7m in Harding's Atlas. Not in Lalande's Catalogue.

[I found it about 7½m and equal to a star preceding it, Nov. 1876.]

(48)

101 Aquarii (b^4).

$23^h. 27^m$, *S* 21°. 35′.

5m Harding and Proctor, 5.4 Heis. 5m Lalande and also 8m., if—as seems to be the case—the star LL 46,126 is identical with b^4. b^4 is possibly variable. [In Sept. 1875. I found it very slightly brighter than 99, but not quite equal to 98.]

7 (Flamsteed) Aquarii.
$20^h. 50^m, S\ 10°. 1$(

This star may possibly be a variable. It was rated 3m (!) by Ptolemy, 6m by Sûfi (10th century), 5m by Argelander, and 6.5 by Heis. In Schjèllerup's translation of Sûfi's Persian manuscript,* its position is thus described :—" La 6ᵉ (7 Fl) est la suivante des trois étoiles situées dans la main gauche, et précède la 4ᵉ (β Aquarii) qui se trouve sur l'epaule gauche ; elle est de la sixième grandeur, tandis que Ptolémée la dit de troisième, mais, en vérité, elle est très—obscure هو خفي جدّا La 7ᵉ (μ Aquarii) se trouve au milieu de ces trois étoiles et précède la 6ᵉ, s'inclinant vers le nord ; elle est des petites de la cinquième grandeur ; Ptolémée la dit de quatrième. Entre elle et la 6ᵉ, il y a environ un empan شبر La 8ᵉ (ι Aquarii) est la précedénte des trois et des grandes de la quatrième grandeur ; Ptolémée la dit de troisième".

The above description clearly proves the star observed by Sûfi to be identical with that numbered 7 by Flamsteed. We cannot, however, be quite so certain that Sûfi rightly identified the star with the one observed by Ptolemy as 3m ; but considering the very careful nature of all

---

* *Description des Étoiles Fixes, composée au milieu du dixième siècle de notre ère par l'astronome Persan Abd-al-rahman al Sûfi. St. Pétersbourgh*, 1874.

Sûfi's descriptions, it seems probable that his identification was correct.

ζ Piscis Australis.

$22^h. 24^m$, S 26°. 41′.

This star was rated 4m by Ptolemy, 5.6 by Sûfi, and 5.6 by Argelander and Heis. It is 6m in Harding's Atlas, but it would seem to have faded of late years. Schönfeld in his second Catalogue of Variable stars* says with reference to this star " U. N. 5.6m. 1864 Aug. und 1869 unter den günstigsten Umständen nur mit Mühe zu sehen, und selbst in dem hochgelegenen Kephissia nur 7m in Opernglas. A. N. 74. 1770"† Auch Heis 5.6m."

ψ³ Aquarii.

$23^h. 13^m$, S 10°. 16′.

With reference to this star Schönfeld remarks " Der südlichste der drei ψ Aquarii ist mir seit Jahren als veränderlich erschienen ; die Periode wird sehr lang sein*. A. N. 77. 1832."

It was rated 4m by Ptolemy and Sûfi, 6m by Lalande, and 5m by Argelander and Heis. It is 5m in Harding's Atlas [I found it brighter than χ Aquarii in Dec. 1875].

* *Zweiter Catalog von Veränderlichen Sternen. Mannheim*, 1875.

† *Astronomische Nachrichten. Vol.* 74, *No.* 1770.